LINCOLN CHRISTIAN COLLEGE AN

P9-DFC-517

Summer for the Gods

Summer for the Gods

The Scopes Trial and America's Continuing Debate over Science and Religion

EDWARD J. LARSON

Harvard University Press
Cambridge, Massachusetts
London, England

Copyright © 1997 by Edward J. Larson
All rights reserved
Printed in the United States of America

First Harvard University Press paperback edition, 1998

Published by arrangement with Basic Books, a division of Perseus Books, LLC.

Designed by Elliott Beard

Library of Congress Cataloging-in-Publication Data

Larson, Edward J. (Edward John)
Summer for the gods : the Scopes trial and America's continuing debate over science and religion / Edward J. Larson — 1st ed.
p. cm.
Includes bibliographical references and index.
ISBN 0-674-85429-2
1. Scopes, John Thomas—Trials, litigation, etc. 2. Evolution—Study and teaching—Law and legislation—United States. I. Title.
KF224.S3L37 1997
345.73′0288—dc21 97-9648

In memory of my father, Rex Larson,
a Darrowlike criminal lawyer

and

William H. Ellis, Jr.,
a Bryanesque attorney-politician

1346

113298

CONTENTS

PREFACE

THE SCOPES TRIAL has dogged me for more than a decade, ever since I wrote my first book on the American controversy over creation and evolution. The trial only constituted one brief episode in the earlier book, yet people who knew of my work asked me more about that one event than everything else in the book combined—and they would tell me about the Scopes trial and what it meant to them. Over the years, their questions and comments led me to reflect on the so-called trial of the century. Finally, one of my colleagues, Peter Hoffer, suggested that I write a separate book solely about the trial and its place in American history. The idea made immediate sense. As a historical event and topic of legend, the trial had taken on a life and meaning of its own independent of the overall creation–evolution controversy. Indeed, this book is different from my earlier one in that they chronicle remarkably separate stories. Both are tales worth telling as stories of our time. Furthermore, no historian had examined the Scopes trial as a separate study in decades. I had access to a wealth of new archival material about the trial not available to earlier historians, and the benefit of additional hindsight.

Many helped me to conceptualize, research, and write this book. A few also assisted me with my first book, particularly my former teachers and current friends Ronald Numbers and David Lindberg. Some I met in the course of my earlier work, such as Bruce Chapman, Richard Cornelius, Edward Davis, Gerald Gunther, Phillip Johnson, William Provine, George Webb, and John West. Others were my colleagues at the University of Georgia, including Betty Jean Craige, Thomas Lessl, Theodore Lewis, William McFeely, Bryant Simon, Phinizy Spalding, Lester Stephens, and Emory Thomas. Finally, I benefited immensely from ongoing advice and encouragement from my editors at Basic Books, Juliana Nocker, Steven Fraser, and Michael Wilde. My thanks go to all of them.

Numerous institutions assisted me by providing research materials and support for this project. Among the sources for research material, I particularly want to acknowledge my debt to the American Civil Liberties Union, Bryan College, the Library of Congress, Princeton University Libraries, the Tennessee State Archives, the University of Tennessee Libraries, and Vanderbilt University Libraries. I owe a special debt to Carolyn Agger for allowing me access to the Fortas Papers. Early and ongoing support came from sources within the University of Georgia, including two Senior Faculty Research Grants from the Vice President for Research; a Humanities Center fellowship; summer support from my dean at the law school, Edward Spurgeon; and travel support from the chair of the history department, David Roberts. The Discovery Institute and the Templeton Foundation provided forums to discuss the ideas that went into this book. Finally, I especially enjoyed an opportunity to work on this project at the Rockefeller Foundation's Bellagio Study Center. This book would not have been possible without such support.

E. L.
Bellagio, Italy
November 1996

SUMMER FOR THE GODS

— INTRODUCTION —

It STARTED OFF civilly enough. Darrow began by asking his world-famous expert witness, "You have given considerable study to the Bible, haven't you, Mr. Bryan?"

"Yes, sir, I have tried to," came the cautious reply.

"Well, we all know you have, we are not going to dispute that at all," Darrow continued. "But you have written and published articles almost weekly, and sometimes have made interpretations of various things?"

Bryan apparently saw the trap. If he assented to having interpreted some biblical passages, then he could scarcely object to others giving an evolutionary interpretation to the Genesis account of human creation. "I would not say interpretations, Mr. Darrow, but comments on the lesson."

The lawyerly game of cat and mouse had begun, but one in which the cat sought to kill his prey and the mouse had nowhere to hide. At 68, Clarence Darrow stood at the height of his powers, America's greatest criminal defense lawyer and champion of anticlericalism. Three years his junior, the former Boy Orator of the Platte—once the nation's youngest major-party presidential nominee and now leader of a fundamentalist crusade against teaching evolution in public schools—William Jennings Bryan remained a formidable stump speaker, although he lacked the quick wit to best Darrow in debate. This was no debate, however; it was a courtroom interrogation in which Darrow enjoyed all the advantages of an attorney questioning a hostile witness.

Although it would become the most famous scene in American legal history, it did not occur in a courtroom. Fears that the huge crowd would collapse the floor forced the judge to move the afternoon's proceedings onto the courthouse lawn, with the antagonists on a crude wooden platform before a sea of spectators, much like Punch and Judy puppets performing at an outdoor festival. Enterprising youngsters passed through the crowd hawking refreshments as Darrow began to question Bryan about various Old Testament miracles.

"Do you believe Joshua made the sun stand still?" Darrow asked at one point, regarding the biblical passage that speaks of a miraculously lengthened day.

"I believe what the Bible says. I suppose you mean that the earth stood still?" Bryan replied, anticipating the standard gibe against biblical literalism under a Copernican cosmology.

Darrow feigned innocence. "I don't know. I am talking about the Bible now."

"I accept the Bible absolutely," Bryan affirmed. "I believe it was inspired by the Almighty, and He may have used language that could be understood at that time instead of using language that could not be understood until Darrow was born."

This rejoinder evoked laughter and applause from the partisan Tennessee audience, yet Darrow had struck a blow; even a biblical literalist such as Bryan recognized the need to interpret some scriptural passages. Darrow drove the point home with further questions. "If the day was lengthened by stopping either the earth or the sun, it must have been the earth?"

"Well, I should say so," an exasperated Bryan sighed, and in so doing fell into another trap.

Darrow snapped it shut by asking, "Now, Mr. Bryan, have you ever pondered what would have happened to the earth if it had stood still?"

"No."

"You have not?" Darrow asked with mock incredulity.

Bryan fell back on faith. "No; the God I believe in could have taken care of that, Mr. Darrow." Now the assembled reporters from across the country smiled among themselves.

"Don't you know it would have been converted into a molten mass of matter?" Darrow asked rhetorically. In giving ground on biblical literalism to accommodate a heliocentric solar system, Bryan fell headlong into problems with terrestrial geology and physics. If he gave any more ground, how then could he hold the line on Genesis? If he had given any less ground, he would have sounded supremely foolish. Yet he had conceded the critical point that scripture required interpretation in light of modern science, and he would do so again with the days of creation and the age of the universe. There were no good answers to these questions from Bryan's perspective.

The chief prosecutor had heard more than enough. For the third time, he tried to stop the exchange. "This is not competent evidence," he objected. But Bryan, serving as special counsel for the state and supposedly assisting the prosecutor, stubbornly clung to the simple wooden chair that served as a makeshift witness stand for the outdoor session. Defense counsel "did not come here to try this case," Bryan shouted back. "They came to try revealed religion. I am here to defend it, and they can ask me any question they please."

The crowd roared its approval. "Great applause from the bleachers," Darrow noted for the record.

"From those whom you call 'yokels,'" Bryan thundered. "Those are the people whom you insult."

Glaring at his adversary, Darrow shot back, "You insult every man of science and learning in the world because he does not believe in your fool religion."

"This has gone beyond the pale of a lawsuit, your honor," the prosecutor pleaded. "I have a public duty to perform, under my oath and I ask the court to stop it." But Bryan would not budge.

The judge deferred to the distinguished witness. "To stop it now would not be just to Mr. Bryan," he ruled. And so it continued, with Darrow inquiring about Noah and the Flood, ancient civilizations, comparative religion, and the age of the earth. Bryan sank deeper into confusion as he struggled to answer the barrage of questions. He affirmed his belief in a worldwide flood that killed all life outside the ark (except perhaps the fish, he tried to joke), but interpreted the six days of creation to symbolize vast periods of time. Yet Darrow never asked about evolution or the special creation of humans in the image of God, questions that Bryan surely would answer with well-honed remarks about the so-called missing link in scientific evidence for human evolution and the profound impact of evolutionary naturalism on public morality and private faith. Like any good trial lawyer—and he was the best—Darrow kept the focus on topics that served his purpose, which did not include giving Bryan a soapbox for his speeches.

As the inquiry departed ever further from any apparent connection to the Tennessee law against teaching evolution supposedly at issue in the trial, the prosecutor objected, "What is the purpose of this examination?"

Darrow answered honestly. "We have the purpose of preventing bigots and ignoramuses from controlling the education of the United States," he declared, "and that is all." That was more than enough, for it justified his efforts to publicly debunk fundamentalist reliance on scripture as a source of knowledge about nature suitable for setting education standards. Darrow had gone to tiny Dayton, Tennessee, for precisely this purpose, with Bryan as his target. Bryan had come to defend the power of local majorities to enact a law—his law—to ban teaching about human evolution in public schools. Two hundred reporters had followed to record the epic encounter. They billed it as "the trial of the century" before it even began. No one cared about the defendant, John Scopes, who had volunteered to test the nation's first antievolution statute. The aged warriors had sparred at a distance for over a week without delivering any decisive blows. Now they went head to head, when Bryan vainly accepted Darrow's challenge to testify to his faith on the witness stand as a Bible expert.

By the end of his two-hour-long ordeal, Bryan seemed intent mainly on redeeming his dignity. "The reason I am answering is not for the benefit of the superior court. It is to keep these gentlemen from saying I was afraid to meet them and let them question me."[1] Yet Bryan knew better than to place himself and his faith in such a vulnerable position. Three years earlier, at the outset of the antievolution crusade, Darrow had asked him similar questions in an open letter to the press. Bryan had ignored them. "Anyone can ask questions, but not every question can be answered. If I am to discuss creation with an atheist, it will be on the condition that we [both] ask questions," he had written about that time. "He may ask the first one if he wishes, but he shall not ask a second one until he answers my first."[2]

Bryan had a long list of ready questions for Darrow and other evolutionists. Chief among them, he would ask about the missing links in the fossil record. "True science is classified knowledge, and nothing therefore can be scientific unless it is true," Bryan was prepared to say in his closing argument, which he planned to give that very day before being waylaid by Darrow. "Evolution is not truth; it is merely a hypothesis—it is millions of guesses strung together. It had not been proven in the days of Darwin; he expressed astonishment that with two or three million species it had been impossible to trace any species to any other species."[3] Where are the missing links? "If evolution be true, they have not found a single link," Bryan had told a Nashville audience while campaigning for the Tennessee antievolution law. More critically, he stressed the missing links between humans and their supposed simian relatives, because the challenged statute only pertained to the teaching of human evolution.[4]

At the time, the popular debate over the status of evolution as science centered largely on the interpretation of fossils. Various types of scientific evidence supported the theory, but short of actually observing the development of new kinds of plants or animals, intermediate fossils linking related species offered the most persuasive "proof" of evolution. Proponents particularly relied on the remarkably complete collection of fossils tracing the development of the American horse over three million years, while opponents harped on the "missing links," especially between humans and other primates. For example, Bryan's chief adversary in the creation–evolution controversy from the scientific viewpoint, American Museum of Natural History president Henry Fairfield Osborn, regularly referred to the equine fossils in his many popular articles, books, and lectures countering the antievolution crusade. "It would not be true to say that the evolution of man rests upon evidence as complete as that of the horse," he conceded in a 1922 exchange with Bryan, but "the very recent discovery of Tertiary man . . . constitutes the most convincing answer to Mr. Bryan's call for more evidence." Tracing humanity's family tree, Osborn added, "Nearer to us is the Piltdown man, found [in] England; still nearer in geologic time is the Heidelberg man, found on the Neckar River; still nearer is the Neanderthal man, whom we know all about. . . . This chain of human ancestors was totally unknown to Darwin. He could not have even dreamed of such a flood of proof and truth."[5]

The expert witnesses summoned by Darrow to Dayton brought this evidence with them, complete with models of the hominid fossils. In the scientific affidavits prepared for the defense, for example, anthropologist Fay-Cooper Cole, geologist Kirtley F. Mather, and zoologist

H. H. Newman detailed hominid development through fossils from Java, Piltdown, Heidelberg, and elsewhere. Although Piltdown man later lost his place in the human family tree, Cole added a new find, "made only a few months ago in Bechuanaland of South Africa," purportedly of a being intermediate between humans and anthropoids. "There is nothing peculiar or exceptional about the fossil record of man. It is considerably less complete than that of the horse, . . . but it is far more complete than that of birds," Newman asserted. "Much has been said by the antievolutionists about the fragmentary nature of the fossil record of man, but many other animals have left traces far less readily deciphered and reconstructed."[6]

Yet Bryan expressed concern only about the teaching of human evolution. "The import of the Tennessee trial is in the presence of Mr. Bryan there," the *Chicago Tribune* warned at the time. "What he wants is that his ideas, his interpretations and beliefs should be made mandatory. When Mr. Darrow talks of bigotry he talks of that. Bigotry seeks to make opinions and beliefs mandatory." Bryan's beliefs did not reject all science, or even all evolutionary theory. "Hands off one thing and one thing alone," the *Tribune* observed, "the divine creation of man, the human being with a soul. You may not teach that the Piltdown man reveals any relationship to the anthropoid ape."[7] Given the preoccupation of both sides with scientific evidence of humanity's anthropoidal ancestry, I begin the story here.

PART I

BEFORE . . .

— CHAPTER ONE —

DIGGING UP CONTROVERSY

AS THE SCIENTIFIC world prepared to celebrate the fiftieth anniversary of Charles Darwin's *Origin of Species* in 1909, an amateur English geologist named Charles Dawson made a momentous find thirty miles from Darwin's country home in southern England. From a laborer digging in a gravel pit on a farm near Piltdown Common, Sussex, Dawson received a small fragment of a human cranium's parietal bone. "It was not until some years later, in the autumn of 1911, on a visit to the spot, that I picked up, among the rain-washed spoilheaps of the gravel-pit, another and larger piece belonging to the frontal region of the same skull," Dawson later reported in an article that shook the scientific world. "As I had examined a cast of the Heidelberg jaw, it occurred to me that the proportions of this skull were similar to those of that specimen."[1] This caught his attention. At the time, the Heidelberg jaw represented one of the only two known fossil remains that scientists then attributed to hominid species ancestral to modern humans. Each of these known remains—the jawbone from near Heidelberg, Germany, and a skullcap, three teeth, and a thighbone discovered in Java—had been found during the preceding two decades and remained the subject of intense scientific controversy. The better-known Neanderthal (or Mousterian) "cave men" contributed little to the story of human evolution because they came from a later era, were fully human, and died out. The Piltdown skull, however, could provide the notorious "missing link" in human evolution.[2]

Dawson now began rummaging through the gravel pit in earnest. After uncovering flint tools and the fossil remains of various prehistoric animals, he took the lot to the paleontologist Arthur Smith Woodward of the British Museum in London. Soon Woodward was in Piltdown with Dawson conducting a systematic excavation of the site. During the summer of 1912, they found more fragments of the Piltdown skull, additional prehistoric animal fossils mixed with human tools, and part of a jawbone with two intact molars. These pieces carried tremendous potential significance. Owing to its size and shape, the cranium clearly came from a hominid. The flint tools reinforced this conclusion. The animal remains and the geology of the site suggested that the skull dated from the Pleistocene epoch, at some point midway between the supposed date of the so-called ape-man of Java and the emergence of modern humans. The jaw, however, appeared to come from a type of ape never known to have lived in Europe, and the teeth were worn down in a human fashion. Pieced together by Woodward, the picture emerged of a new species of extinct hominid that he called *Eoanthropus dawsoni*, or the "dawn man" of Piltdown.

Dawson and Woodward unveiled their discovery on December 18, 1912, before a packed house of Britain's scientific elite at the Geological Society of London. "While the skull, indeed, is essentially human, and approaching the lower grade in certain characters of the brain," they explained at the time, "the mandebile appears to be almost precisely that of an ape, with nothing human except the molar teeth." After describing their find in great detail and fitting it into the sequence of other known fossil remains, Dawson and Woodward concluded, "It tends to support the theory that Mousterian man was a degenerate offshoot of early man, and probably became extinct; while surviving man may have arisen directly from the primitive source of which the Piltdown skull provides the first discovered evidence."[3] Sir Arthur Keith, one of the world's leading experts on human antiquity and anatomy, attended the presentation by Dawson and Woodward, and generally concurred in their conclusions, as did the renowned neurologist Grafton Elliot

Smith and the famed biologist Boyd Dawkins. Perhaps Dawkins best expressed the collective response of the learned audience when he declared during the discussion period, "The evidence was clear that this discovery revealed a missing link between man and the higher apes."[4]

Word of the discovery became front-page news throughout the United States, where prominent creationists still publicly denounced the Darwinian theory of human evolution. Relying on a special same-day cable transcript, the *New York Times* published a summary of Dawson and Woodward's initial presentation within hours of the event. "Paleolithic Skull Is a Missing Link," the *Times* headline proclaimed, "Bones Probably Those of a Direct Ancestor of Modern Man."[5] A day later, the *Times* followed up with a telegraphic interview of Woodward. "Hitherto the nearest approach to a species from which we might have been said to descend that had been discovered was the cave-man," Woodward observed in this interview, "but the authorities constantly asserted that we did not spring direct from the cave-man. Where, then, was the missing link in the chain of our evolution? To me, at any rate, the answer lies in the Piltdown skull, for we came directly from a species almost entirely ape."[6] Other American newspapers carried similar reports.[7]

The *New York Times* concluded its coverage of the Piltdown discovery with an extended, page-one summary of the episode, appearing in its next Sunday edition. "Darwin Theory Is Proved True," proclaimed the banner headline. "English Scientists Say the Skull Found in Sussex Establishes Human Descent from Apes." This article reprinted Keith's observation that the discovery "gives us a stage in the evolution of man which we have only imagined since Darwin propounded the theory."[8] Yet an editorial entitled "Simian Man" appearing in that same Sunday edition cautioned readers, "Those who have read the cable dispatches to The Times describing the oldest human skull . . . must not confuse this ancient man with the 'missing link' or with the ancestry of the present human race. Darwin thought that man was descended from apes, but he searched in vain for the half-man, half-ape." Although the British scientists quoted in those dispatches clearly saw the new fossil as filling a missing link in the record of human evolution, the *Times* editorial cites their classification of the Piltdown hominid as a distinct species to support the conclusion that "he was no forebear of our Adam."[9]

This peculiar editorial disavowing a scientific news report reflected the divided mind of the American public, during the years leading up to the Scopes trial, regarding the controversial topic of human evolution. Of course, no single fossil discovery could prove the Darwinian theory of human evolution. As the *New York Times* editorial suggested, evidence that a "simian man" walked the earth in the Pleistocene epoch does not conclusively establish a simian ancestry for modern humans. Yet it fit into a larger pattern. During the first quarter of the twentieth century, scientists in western Europe and the United States accumulated an increasingly persuasive body of evidence supporting a Darwinian view of human origins, and the American people began to take notice. These scientific developments helped set the stage in the early 1920s for a massive crusade by fundamentalists against teaching evolution in public schools, which culminated in the 1925 trial of John Scopes.

The theory that current living species evolved from preexisting species had been around for a long time. More than a century earlier, a well-known French naturalist, the Chevalier de Lamarck, had proposed a theory of progressive evolutionary development based on vital forces within living things and the inheritance of acquired characteristics. Lamarck viewed the various biological species as arranged in an ascending hierarchy from the simplest to the most complex, reflecting a historical pattern of development. Vital forces within living entities prompted their development, allowing each generation to progress beyond the level of complexity of its ancestors. The use or disuse of organs in response to changed environmental conditions further propelled evolutionary progress, according to Lamarck, as living entities passed their acquired characteristics on to their offspring. The giraffe's neck remains the most famous example of this process. As vegetation became scarcer in their habitat, Lamarck hypothesized, the ancestors of the present-day giraffe stretched their necks to eat the remaining leaves high on trees. The next generation inherited the longer necks and stretched them still further, until a new species of long-neck giraffe evolved.

Although early nineteenth-century scientists generally did not accept Lamarck's ideas on evolution and held to the creationist concept that each biological species remained fixed over time, many of them did embrace the bold theories of Lamarck's rival, Georges Cuvier. As curator of vertebrate fossils at the prestigious French Museum of Natural History, Cuvier was the first Enlightenment-era naturalist forced to come to terms with the increasingly complex fossil record then being unearthed by scientific expeditions. This research drove him to acknowledge that the earth had a very long geologic history, far longer than suggested by a literal interpretation of the account in Genesis, and that countless biological species had appeared and become extinct during that long history, despite the traditional scientific and religious view that all species continued over time. Sudden breaks that appeared in the fossil record in which one characteristic group abruptly replaced an earlier one with few transitional forms, coupled with a conviction that living species were too complex to evolve, led Cuvier to conclude that great catastrophes such as worldwide floods or ice ages punctuated geologic history into a series of distinct epochs. Each catastrophe wiped out most or all living things, leaving the earth to repopulate through migration by the few survivors, as Cuvier at first supposed, or new creations of biological species, as later naturalists concluded after wider exploration found no ancient source for modern animals.

Cuvier's theories quickly came to dominate the geological thinking of the day. Some secular scientists in that era of romanticism and transcendentalism attributed the successive new creations of species to a vital force within nature. Christian geologists, in contrast, saw the hand of God directly at work in these creative acts. Both groups, however, accepted a long geologic history and the progressive appearance of new life forms. For Christians, this posed a conflict with the account in Genesis, which declared that God formed the heavens, the earth, and all kinds of living things in six days, culminating in the creation of Adam and Eve as the forebears of all human beings. In the fifteenth century, the scholarly archbishop James Ussher used internal evidence within Genesis to fix the year of creation at 4004 B.C. Even if they did not adopt this precise year, many later Christians accepted a similar time frame for the creation. In America during the middle part of the nineteenth century, such leading geologists as Amherst College president Edward Hitchcock and Yale's James D. Dana reconciled contemporary geological opinion with their traditional religious beliefs by interpreting the biblical days of creation as symbolizing geologic ages or, alternatively, by positing a gap in the Genesis account.[10]

Nineteenth-century Protestants, including many with decidedly conservative views of scriptural authority, readily accepted such accommodations of science and religion. Even the *Scofield Reference Bible*, which profoundly influenced the development of modern fundamentalism around the turn of the twentieth century, incorporated the "gap theory" into its explanation of Genesis and referred to the "day/age theory" in a footnote.[11]

The advent of Darwinism presented a far greater threat to Christians than simply a long geological history and the progressive appearance of species. When Darwin's *Origin of Species* first appeared in 1859, few scientists accepted the concept of organic evolution. Within two decades, however, even a hostile church journal could identify only two working American naturalists who still opposed it.[12] Darwin's eloquent presentation of evidence for evolutionary development drawn from careful observation of nature certainly contributed to this turnabout, but he proposed also that a "survival of the fittest" process of natural selection drove evolutionary change rather than the benign process of individual adaptation envisioned by Lamarck. Although Darwin always maintained a place for Lamarckian-type mechanisms within his theory of evolution, his concept of natural selection became widely identified as the central feature of Darwinism.

The high school textbook at issue in the Scopes trial, George William Hunter's *A Civic Biology*, summarized Darwin's alternative evolutionary mechanism in a section entitled "Charles Darwin and Natural Selection." Darwin observed that individual plants and animals tended to vary slightly from their ancestors, Hunter noted. "In nature, the variations which best fitted a plant or animal for life in its environment were the ones which were handed down because those having variations which were not fitted for life in that particular environment would die," Hunter wrote. "Thus nature seized upon favorable variations and after a time, as the descendants of each of these individuals also tended to vary, a new species of plant or animal, fitted for the place it had to live in, would be gradually evolved." In short, as Hunter explained, Darwin postulated new species "arising from very slight variations, continuing during long periods of years."[13] This mechanism attributed these all-important variations to random individual differences inborn in the offspring rather than to Lamarckian vital forces or acquired characteristics. "Species have been modified, during a long course of descent," Darwin concluded in the *Origin of Species,* "chiefly through the natural selection of numerous successive, slight, favorable variations."[14]

Darwin's account of random variations, coupled with his survival-of-the-fittest selection process, posed a critical problem for many Christians who retained a teleological view of nature. In 1860, Darwin anticipated this problem in an exchange with the Harvard botanist Asa Gray, a devout Protestant. Christians long maintained that the harmonious structure of the physical universe and each living thing reflected intelligent design by a creator, and thereby contributed evidence of the existence and loving character of God. Gray, who had arranged the initial publication of the *Origin of Species* in the United States, asked Darwin about the book's theological implications. "I had no intention to write atheistically," Darwin replied. "But I own that I cannot see as plainly as others do, and as I should wish to do, evidence of design and beneficence on all sides of us. There seems to me too much misery in the world. I cannot persuade myself that a beneficent and omnipotent God would have designedly created the Ichneumonidae with the express intention of their feeding within the living bodies of Caterpillars, or that a cat should play with mice."[15] For some conservative theologians and pious scientists, this represented the ultimate challenge of Darwinism to a Christian world view: Beneficial variation was random and natural selection was cruel. If nature reflected the character of its creator, then the God of a Darwinian world acted randomly and cruelly. Darwin could not accept such a God, and became an agnostic. Others recognized the magnitude of the issue.

Battle lines formed quickly. The self-proclaimed "gladiator-general" of Darwinism, English naturalist T. H. Huxley, claimed to take up the banner for science.[16] Anticipating religious opposition to Darwin's ideas, the agnostic Huxley—who embraced the theory of evolution as a naturalistic rebuttal to the claims of Christianity—wrote to Darwin shortly before publication of *Origin of Species,* "I am sharpening up my claws and beak in readiness."[17] Following publication, Huxley aggressively championed the cause in countless public debates and popular articles, clashing with such religiously motivated critics of Darwinism as Oxford bishop Samuel Wilberforce and British prime minister William Gladstone. "Whether astronomy and geology can or cannot be made to agree with . . . Genesis," Huxley wrote in a typical passage, "are matters of comparatively small moment in the face of the impassable gulf between the anthropomorphism (however refined) of theology and the passionless impersonality . . . which science shows everywhere underlying the thin veil of phenomena. Here seems to me to be the great gulf fixed between science and theology."[18] Counterattacking in the name of religion, Princeton theologian Charles Hodge took the lead in challenging Darwinism. His provocatively titled 1874 book *What is Darwinism?* presented a tightly reasoned argument that led to the inevitable answer: "It is atheism [and] utterly inconsistent with the Scriptures." For Hodge, Darwin's "denial of design in nature is virtually the denial of God."[19]

Hodge and some other church leaders raised an alarm against teaching evolution, particularly within seminaries and denominational colleges, but scientific developments temporarily quieted the conflict. In the 1870s and 1880s, Darwinism faced a host of technical challenges. The best evidence from the physical sciences suggested that the solar system was not old enough for slight, random variations in one or more organisms to produce the current array of biological species, much less to generate life from nonlife. Further, without a means to preserve inherited differences, such variations did not lead anywhere. Like most naturalists working before the acceptance of Mendelian genetics, Darwin believed that the inherited traits of an offspring consisted of a blending of those possessed by its parents. Slight, random variation in an individual—no matter how much it helped that animal or plant survive—quickly would be swamped as that individual bred with others of its species, so that gradually each succeeding generation would lose its distinctiveness. Even if individuals with a particularly beneficial trait mated solely with those possessing the same trait—such as happens in the breeding of domesticated animals—their offspring then simply would tend to preserve that trait, not exceed it. If organic evolution occurred (and by 1880 most naturalists believed that it did), then some mechanism must accelerate and direct variation; for some devout Christians, this left a role for God.

Two alternative theories of evolution were discussed widely among American and European scientists during the final third of the nineteenth century. Ever the traditional Christian, Asa Gray proposed a theory of theistic evolution in which God channeled variations into a pattern of progressive development. The renowned British scientists Charles Lyell, Richard Owen, and St. George Mivart toyed with similar ideas. For some, this offered a way to reconcile religious faith with evolutionary theory and science. Other naturalists, led in the United States by the likes of Joseph LeConte, Clarence King, and Edward Drinker Cope, revived Lamarckian-type explanations to account for the speed and direction of evolution. According to these late-nineteenth-century naturalists—some of whom went so far as to call themselves "neo-Lamarckians"—indwelling vital forces pulled each species forward toward increasing complexity, while each individual pushed in the same direction through the use and disuse of organs in response to shared environmental conditions. Variations became purposeful and natural selection marginalized.

These alternative theories of evolution might not fit neatly with traditional Christian doctrines, but they certainly could be spiritual. In a lecture to Yale seminarians, for example, Gray declared that with evolution, "the forms and species, in all their variety, are not mere ends in themselves, but the whole a series of means and ends, in the contemplation of which we may obtain higher and more comprehensive, and perhaps worthier, as well as more consistent, views of design in Nature than heretofore."[20] Similarly, the neo-Lamarckians' principal journal, *American Naturalist,* professed a goal of "illustrating the wisdom and goodness of the Creator."[21] LeConte defined the "laws of evolution" as "nought else than the mode of operation of the . . . divine energy in originating and developing the cosmos."[22] King denounced natural selection: "A mere Malthusian struggle was not the author and finisher of evolution; but that He who brought to bear that mysterious energy we call life upon primeval matter bestowed at the same time a power of development by change."[23] A Quaker turned Unitarian, Cope concluded in his *Theology of Evolution,* "The Neo-Lamarckian philosophy is entirely subversive to atheism."[24] Conservative Christians might disagree with these views on various points of doctrine, but few raised loud objections, and many liberal Christians wholly embraced an evolutionary creed.[25]

Neo-Lamarckianism and other non-Darwinian forms of evolutionary thought swept the scientific community, particularly in the United States. "From the high point of the 1870s and 1880s, when 'Darwinism' had become virtually synonymous with evolution itself, the selection theory had slipped in popularity to such an extent that by 1900 its opponents were convinced it would never recover," the historian Peter J.

Bowler observed. "Evolution itself remained unquestioned, but an increasing number of biologists preferred mechanisms other than selection to explain *how* it occurred."[26] Even Darwin granted an ever larger role to Lamarckian explanations for variation in later editions of the *Origin of Species*. "The fair truth is that Darwinian selection theories," Stanford zoologist Vernon L. Kellogg concluded in 1907, "stand to-day seriously discredited in the biological world."[27]

With the "eclipse of Darwinism," as T. H. Huxley's grandson Julian later referred to this period in the history of biology, many conservative Christians toned down their rhetoric. "I do not carry the doctrine of evolution as truth as some do," William Jennings Bryan assured audiences around the turn of the century. But he quickly added, "I do not mean to find fault with you if you want to accept the theory; all I mean to say is that while you may trace your ancestry back to the monkey if you find pleasure or pride in doing so, you shall not connect me with your family tree without more evidence than has yet been provided."[28] Apparently the evidence satisfied such highly orthodox Protestant theologians as Princeton's James McCosh and Rochester seminary president A. H. Strong, who now took the position that Christians could accept evolution as, to use Strong's words, "the method of divine intelligence" in creation.[29]

A similarly conciliatory tone sounded in some early essays in *The Fundamentals*, a series of popular booklets published between 1905 and 1915 that helped define the tenets of Protestant fundamentalism. Princeton theologian B. B. Warfield contributed an article to the first volume of this series about the same time as he publicly endorsed theistic evolution as a tenable theory of the "divine procedure in creating man."[30] The theologian James Orr allowed his favorable views on organic evolution to spill over into his four essays for *The Fundamentals*. Earlier he had written, "Assume God—as many devout evolutionists do—to be immanent in the evolutionary process, and His intelligence and purpose to be expressed in it; then evolution, so far from conflicting with theism, may become a new and heightened form of the theistic argument."[31] In *The Fundamentals*, Orr added, "Much of the difficulty on this subject has arisen from the unwarrantable confusion or identification of evolution with *Darwinism*." Now that the "*insufficiency of 'natural selection'*" has been widely recognized by scientists, Orr asserted that evolution was "coming to be recognized as but a new name for 'creation.'" Based on this endorsement for theistic evolution, Orr could confidently proclaim, "Here, again, the Bible and science are felt in harmony."[32]

By the turn of the century, secular historians and essayists rather than theologians and scientists were largely responsible for keeping alive the public perception of hostility between Christians and evolutionists. During the last third of the nineteenth century, two academicians from New York, John William Draper and Andrew Dickson White, wrote enormously popular but highly biased histories of the relationship between science and religion. Draper described his *History of the Conflict Between Religion and Science* as "a narrative of the conflict of two contending powers, the expansive force of the human intellect on one side, and the compression arising from traditionary faith and human interests on the other."[33] White opened his *Warfare of Science* with the sentence, "I propose to present an outline of the great, sacred struggle for the liberty of science—a struggle which has lasted for so many centuries, and which yet continues."[34] He later fleshed out this brief book into a massive, two-volume *A History of the Warfare of Science with Theology in Christendom*. These books recounted Roman Catholic attacks on Copernican astronomy, including the seventeenth-century trial of Galileo and execution of Giordano Bruno, and fostered the impression that religious critics of Darwinism threatened to rekindle the Inquisition. They neither reported the growing harmony between theologians and evolutionists nor noted that most great physical scientists of the period, from John Dalton and Michael Faraday to Lord Kelvin and James Clerk Maxwell, were devout Christians. Instead, as James Orr complained in *The Fundamentals* about these books, "Science and Christianity are pitted against each other. Their interests are held to be antagonistic."[35]

This contentious view of science and religion gained a wide following among secular scholars during the early twentieth century and stiffened their resolve to defy Bryan's antievolution crusade during the 1920s. "Andrew D. White's *Warfare of Science with Theology* is responsible for much of their thinking about religious bigotry and intolerance, and they are ready to join in smiting the Infamous," famed Vanderbilt University humanist Edwin Mims observed of his fellow academics in an address to the Association of American Colleges in 1924. "In other words, college professors are like most human beings in not being able to react to one extreme without going to the other."[36] During the years leading up to the Scopes trial, this reaction inspired an outpouring of academic books, articles, and essays discussing the conflict between science and religion, with an increasing focus on the seemingly pivotal issue of Darwinism. During the first decade of the century, for example, one commentator wrote that Darwin's theory "seemed to promise the greatest victory ever yet won by science over theology." To another, it "constituted the final and irresistible onslaught of science on the old view as to the nature of Biblical authority."[37] In 1922, Piltdown fossil expert Arthur Keith wrote of Darwin and Huxley, "They made it possible for us men of to-day to pursue our studies without persecution—without being subject to the contumely of Church dignitaries."[38]

By 1925, the warfare model of science and religion had become ingrained into the received wisdom of many secular Americans. Clarence Darrow imbibed it as a child in Kinsman, Ohio, where his fiercely anticlerical father eagerly read Draper, Huxley, and Darwin, and made sure that his son did too.[39] As a Chicago lawyer and politician in the 1890s, Darrow quoted Draper and White in his public addresses and denounced Christianity as a "slave religion" that "sought to strangle heresy by building fires around heretics."[40] Similar views characterized Scopes's other defenders. For example, en route to Dayton, defense cocounsel Arthur Garfield Hays told reporters, "Of all the books I have read for this trial, the 'Warfare Between Science and Religion [*sic*],' by Prof. White, is, to my mind, one of the most interesting and readable." He quoted from this book in the course of his legal argument in Dayton and distributed it to at least some of the people that he met there.[41] The zoologist Winterton C. Curtis, who served as an expert witness for Scopes, did not need a copy—he knew the story by heart. "I remembered how, as a college student in the mid-nineties, I had almost wished that I had been born twenty years earlier and had participated in the Thirty Year War [between Darwinists and Christians], when the fighting was really hot," Curtis later recalled. "When, in the second decade of the present century, some of my former students, who had become teachers, began to report the restrictions laid upon them in high schools and in some denominational colleges, I . . . [assumed] an active part in the defense of evolution."[42]

As Curtis suggested, the warfare between fundamentalists and evolutionists revived by the 1920s, along with the fortunes of Darwinism.

Darwin historian James R. Moore described this renewed controversy: "Fifty years it had taken for the teaching of evolution to filter into the high schools, for the high schools to reach the people, and for the people—those, at any rate, who became militant Fundamentalists—to belong to a generation who could not remember the evangelical evolutionists among its ancestors."[43] Moore identified two different causes for the timing of the antievolution crusade here. First, Darwinism did not become a fighting matter for many fundamentalists until it began to influence their children's education in the twenties. Second, Christian biologists at that time could not so readily step in, as they had earlier, to soften evolution's impact on religious belief. Largely due to developments in experimental genetics, biologists increasingly accepted random, inborn variation as the driving force for evolutionary change and rejected the Lamarckian-type explanations that diminished the role of natural selection. Both were significant causes.

Evolutionary theory did not suddenly appear in American high school education at the time of the antievolution crusade; it had been incorporated into leading textbooks during the late nineteenth century, but with a theistic or Lamarckian twist that reflected prevailing scientific opinion. Asa Gray's popular text, for example, explained how evolutionary relationships showed that biological species "are all part of one system, realizations in nature, as we may affirm, of the conception of One Mind."[44] Joseph LeConte organized his 1884 high school textbook around the concept of evolution without ever mentioning natural selection. Purposeful non-Darwinian mechanisms dispensed with the need for chance variations and a naturalistic struggle for survival.[45]

Textbooks typically became more Darwinian in the new century, however, especially after the newly organized field of biology began to replace separate courses on botany and zoology in the high school curriculum. One representative biology text featured a picture of Darwin and a subchapter titled "The Struggle for Existence and Its Effects."[46] Another hailed Darwin for discovering "the laws of life," including the concept of organic evolution through natural selection.[47] Hunter's *Civic Biology*, the best-selling text in the field, credited Darwin for "the proofs of the theory on which we to-day base the progress of the world." This view of progress was decidedly anthropocentric and heavily laced with the scientific racism of the day. According to Hunter, "simple forms of life on the earth slowly and gradually gave rise to those more complex." Humans appeared as a progressive result of this evolutionary process, with the Caucasian race being "finally, the highest type of all."[48] Overall, Darwinism did not feature prominently in Hunter's books or in other early twentieth century biology texts that stressed practical problems, but the concept of organic evolution pervaded the whole of them.

Darwinian concepts in public secondary education touched more families, and more fundamentalists, as the new century unfolded. Relatively few American teenagers attended high school during the nineteenth century, and nearly none did so in the rural South, where such schools rarely existed and local authorities did not compel student attendance. The situation changed dramatically after the turn of the century. Census figures tell the story. The number of pupils enrolled in American high schools lept from about 200,000 in 1890, when the federal government began collecting these figures, to nearly two million in 1920. Tennessee followed this national trend, with its high school population rising from less than 10,000 in 1910 to more than 50,000 at the time of the Scopes trial in 1925. This increase resulted in part from tougher Progressive-era school attendance laws that forced more teenagers to go to school, and followed also from greater access to secondary education, as the number of public high schools increased dramatically during the early part of the century.[49] Commenting on this trend with respect to Tennessee, Governor Austin Peay—who signed the state's antievolution bill into law—boasted in his 1925 inaugural address, "High schools have sprung up throughout the state which are the pride of their communities."[50] This was certainly true for Dayton, site of that year's Scopes trial, which opened its first public high school in 1906.[51] These new schools inevitably included Darwinian concepts in their biological classes, in line with modern developments in American scientific thought.

Hunter's *Civic Biology* reflected some of these scientific developments by including sections on both natural selection and genetics. In designing the new biology curriculum for secondary schools, Hunter and his colleagues at New York's DeWitt Clinton High School worked closely with educators at nearby Columbia University. The Columbia faculty included many leading educators at the university's famed Teachers College and America's foremost geneticist, Thomas Hunt Morgan. While Hunter sought the advice of education experts in shaping the contents of biology instruction, one of his closest colleagues earned a doctorate under Morgan, then in the process of laying the foundations of modern genetics.

Morgan began his groundbreaking research at the turn of the century as an opponent of both gradual Darwinian natural selection and static Mendelian genetics. He favored an alternative theory of rapid evolution by the occurrence and hereditary transmission of inborn mutations. Through experiments with generations of fruit flies, Morgan came to recognize that the inheritance of mutations followed a Mendelian pattern that could provide the basis for a Darwinian form of evolution. Under Mendelianism, he reasoned, even slight mutations in an individual plant or animal would survive in the population and could succeed by means of natural selection. "Evolution has taken place by the incorporation into the race of those mutations that are beneficial to the life and reproduction of the organism," Morgan wrote in 1916. "Natural selection as here defined means both the increase in the number of individuals that results after a beneficial mutation has occurred (owing to the ability of living matter to propagate) and also that this preponderance of certain kinds of individuals in the population makes some further results [in the same direction] more probable than others."[52]

Morgan never fully accepted the sufficiency of slight, random variations to account for the emergence of new species. He continued to rely on mutations to fuel evolution, with natural selection acting as a sieve, and rejected, as he later wrote, "Darwin's postulate that the individual variations, everywhere present, furnished the raw material for evolution."[53] It took a generation of research by population geneticists, biometricians, traditional Mendelianists, and field naturalists to construct the modern neo-Darwinian synthesis that today dominates scientific thought. By the 1920s, however, the Darwinian mechanisms of random variation and natural selection were returning to center stage in biology.[54] Most fundamentalists never recognized these subtle developments within evolutionary theory and simply rejected all forms of evolution as contrary to a literal reading of scripture, yet for conservative Christians troubled more by the implications of random variation and natural selection than by the general concept of organic evolution, and Bryan fell into this camp, the ground for accommodation was shrinking. And everyone engaged with the issue could understand such bold proclamations as those of the popular science writer A. E.

Wiggam, who commented on the Scopes trial, "Mr. Bryan did not even know that evolution takes place . . . in the hereditary units called 'genes.' . . . Morgan and his students . . . have adduced evidence that these genes are themselves the subject of change. And if these genes can be proved to change . . . then, the case for evolution is absolutely won."[55]

As the example of Morgan illustrates, Darwinism revived gradually. Biologists continued to defend a variety of evolutionary mechanisms, including Lamarckian ones, for a generation; the modern neo-Darwinian synthesis did not fully emerge until the 1940s, but the lack of consensus simply emboldened the antievolution crusaders. Bryan and some other crusader leaders mastered the technique of using scientific arguments against Darwinian mechanisms to attack the theory of organic evolution, infuriating evolutionary biologists. After Bryan asserted in a 1922 essay published by the *New York Times* that "natural selection is being increasingly discredited by scientists,"[56] American Museum of Natural History president Henry Fairfield Osborn, a renowned paleontologist and science popularizer, demanded equal time. "I am deeply impressed with the fact that he has familiarized himself with many of the debatable points in Darwin's opinions," Osborn offered. "Mr. Bryan, who is an experienced politician, and who has known politicians to disagree, should not be surprised or misled when naturalists disagree in matters of opinion. No living naturalist, however, so far as I know, differs as to the immutable truth of evolution . . . of all the extinct and existing forms of life, including man, from an original and single cellular state."[57]

In a similar appeal to the public, Morgan observed, "It is the uncertainty concerning the factors of evolution that has given the opponents of the theory of evolution an opportunity to attack the theory itself." He characterized natural selection as "a theory within a theory" of evolution. "It is an easy task," Morgan warned, "for the anti-evolutionists, by pointing out the conflict of opinion concerning the *causes* of evolution, to confuse this issue with that involving only the interpretation of the factual evidence showing that evolution has taken place."[58] Neither Osborn nor Morgan accepted natural selection of slight, random variations as the sole mechanism for evolution, but both took their stand against the antievolutionist crusade.

A further "scientific" development spurred Bryan and some other antievolutionists. Many Americans associated Darwinian natural selection, as it applied to people, with a survival-of-the-fittest mentality that justified laissez-faire capitalism, imperialism, and militarism. Decades before the crusade, for example, Andrew Carnegie and John D. Rockefeller, Sr., claimed this as justification for their cutthroat business practices. Bryan, who built his political career on denouncing the excesses of capitalism and militarism, dismissed Darwinism in 1904 as "the merciless law by which the strong crowd out and kill off the weak."[59] During the years immediately preceding the antievolution crusade, a scientific-sounding form of these social doctrines gained widespread public attention under the name *eugenics*. In one of his popular textbooks, Hunter defined this term as "the science of improving the human race by better heredity."[60] This new "science" was first proposed by Darwin's cousin, the English scholar Francis Galton, in the 1860s as a means to accelerate beneficial human evolution. The idea attracted few supporters until the turn of the century, when developments in Mendelian genetics made it appear plausible. British eugenicists always associated their cause with Darwin, especially after Darwin's son Leonard assumed presidency of the national Eugenics Education Society. Hence in England, for example, a passion to prove eugenics inspired the evolutionary biologist Ronald A. Fisher to pursue research that, beginning in 1918, helped establish the modern neo-Darwinian synthesis.

In America, many evolutionary biologists embraced eugenics early in the century, but the public campaign to impose eugenic restrictions on reproduction peaked in the twenties. As a result, the eugenics movement coincided with the antievolution crusade in many states. Typically justifying their actions on the basis of evolutionary biology and genetics, by 1935, thirty-five states enacted laws to compel the sexual segregation and sterilization of certain persons viewed as eugenically unfit, particularly the mentally ill and retarded, habitual criminals, and epileptics. "If such people were lower animals, we would probably kill them off to prevent them from spreading," Hunter explained in his *Civic Biology*. "Humanity will not allow this, but we do have the remedy of separating the sexes in asylums or other places and in various ways preventing intermarriage and the possibility of perpetuating such a low and degenerate race."[61]

Some antievolutionists denounced eugenics as the damnable consequence of Darwinian thinking: First assume that humans evolved from beasts and then breed them like cattle. Bryan decried the entire program as "brutal" and at Dayton offered it as a reason for not teaching evolution.[62] Everywhere the public debate over eugenics colored people's thinking about the theory of human evolution. Popular evangelist Billy Sunday, for example, repeatedly linked eugenics with teaching evolution during his 1925 Memphis crusade, which coincided with legislative consideration of the Tennessee antievolution bill. "Let your scientific consolation enter a room where the mother has lost her child. Try your doctrine of the survival of the fittest," Sunday proclaimed at one point. "And when you have gotten through with your scientific, philosophical, psychological, eugenic, social service, evolution, protoplasm and fortuitous concurrence of atoms, if she is not crazed by it, I will go to her and after one-half hour of prayer and the reading of the Scripture promises, the tears will be wiped away."[63] Such prominent eugenicists as A. E. Wiggam recognized a tie between antievolutionism and opposition to eugenics. At the outset of the antievolution crusade, he criticized Sunday and Bryan for not supporting eugenics.[64] Later on, he lamented that "until we can convince the common man of the *fact* of evolution . . . I fear we cannot convince him of the profound ethical and religious significance of the thing we call eugenics."[65]

As much as fundamentalists deplored the social and religious consequences, however, the scientific evidence for human evolution kept accumulating. Late in the summer of 1924, a South African university student brought a fossilized skull to her anatomy professor, Raymond A. Dart. He identified the skull as coming from an ancient baboon and was fascinated by the round hole in its braincase. He promptly sought more specimens from the source of the find, a limestone quarry at Taungs. Two crates of fossils arrived later that fall. "As soon as I removed the lid a thrill of excitement shot through me. On the very top of the rock heap was what was undoubtedly an endocranial cast or mold of the interior of a skull," Dart later recalled. "I knew at a glance that what lay in my hand was no ordinary anthropoidal brain."[66]

Dart rushed into print with his discovery. "Unlike [Java man], it does not represent an ape-like man, a caricature of a precocious hominid failure, but a creature well advanced beyond modern anthropoids in just those characteristics, facial and cerebral, which are to be anticipated in an extinct link between man and his simian ancestor,"

Dart announced in a February 1925 scientific publication. "At the same time, it is equally evident that a creature with anthropoid brain capacity . . . is no true man. It is therefore logically regarded as a man-like ape."[67] The Scottish anthropologist Robert Broon noted, "If an attempt is made to reconstruct the adult skull it is surprising how near it appears to come to [Java man]—differing only in the somewhat smaller brain and less erect attitude. While nearer to the anthropoid ape than man, it seems to be the forerunner of such a type as [Piltdown man], which may be regarded as the earliest human variety."[68] Arthur Keith was more cautious, to which Dart replied, "If any errors have been made they are all on the conservative (ape) side and it is certain that subsequent work will serve only to emphasise the *human* characteristics."[69]

Dart identified one particular human characteristic of the Taungs man-ape that would especially trouble Bryan and the antievolutionists: In trying to deduce how the creature could have survived on the dry plains of the Transvaal, Dart remembered the round hole in the baboon skull from the same site. "Was it possible that the opening had been made by another creature to extract its brain for food?" he asked himself. "Did this ape with the big brain catch and eat baboons? If so it must have been very clever."[70] Such reasoning crept into Dart's initial publication. Paleontologists had mistakenly looked for the missing link in "the luxuriant forests of the tropical belts," he wrote. "For the production of man a different apprenticeship was needed to sharpen the wits and quicken the higher manifestations of intellect. . . . Southern Africa, by providing a vast open country with occasional wooded belts and a relative scarcity of water, together with a fierce and bitter mammalian competition, furnished a laboratory such as was essential to this penultimate phase of human evolution."[71] In short, humans evolved through hunting. As Bryan had warned, Darwin's dreadful law of hate was replacing the Bible's divine law of love as the origin of humanity.[72]

The Johannesburg *Star* scooped the story four days before Dart's official announcement. The news spread fast. A front-page article in the next morning's edition of the *New York Times* proclaimed, "New-Found Fossilized Skull May Be That of Missing Link." Other newspapers followed suit. A popular magazine removed God from the picture altogether in its poetic rendition of events:

Here lies a man, who was an ape.
Nature, grown weary of his shape,
Conceived and carried out the plan
By which the ape is now the man.

Many conservative Christians were openly hostile. A letter to the editor in the London *Times* appealed to Dart: "Man stop and think. You . . . have become one of the Devil's best arguments in sending souls to grope in the darkness." Bryan dismissed all the fossil remains of early humanoids as inconclusive and inconsequential. Many other antievolutionists did the same. Less than two months after Dart's announcement, a New York newspaper reported, "Professor Dart's theory that the Taungs skull is a missing link has evidently not convinced the legislature of Tennessee, the governor of which state has signed an 'Anti-Evolution' Bill which forbids the teaching . . . that man is descended from lower order of animals."[73] Plaster models of the Taungs skull and Piltdown fossils soon appeared as evidence for the defense in Scopes's legal challenge to that new law.

CHAPTER TWO

GOVERNMENT BY THE PEOPLE

FOSSIL DISCOVERIES provided persuasive new evidence for human evolution and as such provoked a response from antievolutionists. Henry Fairfield Osborn threw down the gauntlet in his reply to Bryan's 1922 plea in the *New York Times* for restrictions on teaching evolution. Bryan had argued that "neither Darwin nor his supporters have been able to find a fact in the universe to support their hypothesis,"[1] prompting Osborn to cite "the Piltdown man" and other recent hominid fossil finds. "All this evidence is today within reach of every schoolboy," Osborn wrote. "It will, we are convinced, satisfactorily answer in the negative [Bryan's] question, 'Is it not more rational to believe in the creation of man by separate act of God than to believe in evolution without a particle of evidence?' "[2] Of course, the fact that all this evidence *was* within the reach of every public-school student constituted the nub of Bryan's concern, and Osborn further baited antievolutionists by stressing how it undermined belief in the special creation of humans.

During the years leading up to the Scopes trial, antievolutionists responded to such evidence in various ways. The fundamentalist leader and Scopes trial consultant John Roach Straton, for example, denounced Piltdown man as a fraud.[3] The adventist science educator George McCready Price, who devised a creationist theory of geologic history that Bryan cited at trial, challenged the antiquity and evolutionary order given to the fossilized humanoids. Placing their age at only a few thousand years rather than the hundreds of thousands of years reckoned by Osborn, Price wrote in 1924, "Such specimens as those from Heidelberg, Neanderthal, and Piltdown may be regarded as degenerate offshoots which had separated from the main stock both ethnically and geographically."[4] Bryan simply ridiculed paleontologists. "The evolutionists have attempted to prove by circumstantial evidence (resemblances) that man is descended from the brute," he declared in a 1923 address to the West Virginia state legislature. "If they find a stray tooth in a gravel pit, they hold a conclave and fashion a creature such as they suppose the possessor of the tooth to have been, and then they shout derisively at Moses." Responding in kind, Bryan then shouted derisively at people like Osborn: "Men who would not cross the street to save a soul have traveled across the world in search of skeletons."[5]

The tone of these comments reflected the newfound militancy that characterized the conservative Christians from various Protestant denominations who called themselves fundamentalists during the 1920s and drew together to support the prosecution of John Scopes. Certainly some conservative Christians rejected Darwinism all along, but when doing so even Bryan earlier had added, "I do not mean to find fault with you if you want to accept the theory."[6] Some articles in *The Fundamentals* dating from 1905 to 1915 criticized the theory of evolution, but others in that series accepted it. Indeed, the Baptist leader who founded the series and later helped launch the fundamentalist movement, A. C. Dixon, once expressed his willingness to accept the theory "if proved," while a subsequent series editor, R. A. Torrey, persistently maintained that a Christian could "believe thoroughly in the absolute infallibility of the Bible and still be an evolutionist of a certain type."[7] Such tolerance largely disappeared during and after the First World War, as the fundamentalist movement coalesced out of various conservative Christian traditions.

Militant antievolutionism had not marked any of the four strands of nineteenth-century Christian theology that more or less came together under the fundamentalist banner during the 1920s, yet each joined in the new crusade against teaching evolution. Dispensational premillennialists such as Baptist leaders Dixon, Torrey, and C. I. Scofield brought an intellectual tradition of rigid biblical interpretation that divided history into separate divine dispensations and eagerly anticipated Christ's second coming to replace the current fallen age with a new millennium of peace and justice. Although their otherworldly faith pulled them away from political activism, their biblical literalism committed them to defend the Genesis account of creation. Conservative theologians at the Presbyterian seminary in Princeton added a formal theory of biblical inerrancy, leading their denomination to adopt a five-point declaration of essential doctrines that became central tenets of fundamentalism: the absolute accuracy and divine inspiration of scripture, the virgin birth of Christ, salvation solely through Christ's sacrifice, the bodily resurrection of Christ and his followers, and the authenticity of biblical miracles. Even though at least one founder of this school, the Princetonian B. B. Warfield, accepted theistic evolution, it clearly inclined followers toward a literal interpretation of Genesis.

The two other strands feeding into fundamentalism contributed to the cause more in terms of numbers than doctrines. The holiness movement, which grew out of Methodism to form a variety of small Protestant denominations, certainly clung to the Bible as true, but stressed personal piety and Christian service over intellectual issues. Penticostalism, which was then entering a period of dramatic growth that would last throughout the century, built on solid premillennialist and holiness foundations, but set them holy rolling by emphasizing the miraculous work of the Holy Spirit in the lives of individual believers. Both groups brought to the antievolution crusade an army of loyal foot soldiers ready to fight any public-school teachings that threatened to undermine the religious faith of their children. Bryan, a practical politician with great personal faith in the Bible and no formal theological training, did not fit neatly into any one of these camps, but shared with them a sense that something was wrong with mainline Protestantism and American culture.

The culprit, they all agreed, was a form of theological liberalism known as "modernism" that was gaining acceptance within most mainline Protestant denominations. Modernists viewed their creed as a means to save Christianity from irrelevancy in the face of recent developments in literary higher criticism and evolutionary thinking in the social sciences. Higher criticism, especially as applied by German theologians, subjected the Bible to the same sort of literary analysis as any other religious text, interpreting its "truths" in light of its historical and cultural context. The new social sciences, particularly psychology and anthropology, assumed that Judaism and Christianity were natural developments in the social evolution of the Hebrew people. Modernists responded to these intellectual developments by viewing God as immanent in history. Conceding human (rather than divine) authorship for scripture and evolutionary development (rather than revelational truth) for Christianity, modernists nevertheless claimed that the Bible represented valid human perceptions of how God acted. Under this view, the precise historical and scientific accuracy of scripture did not matter. Judeo–Christian ethical teachings and individual religious sentiments could still be "true" in a realm beyond the "facts" of history and science. "In brief," the modernist leader Shailer Mathews of the University of Chicago divinity school wrote in 1924, "the use of scientific, historical, and social methods in understanding and applying evangelical Christianity to the needs of living persons, is Modernism."[8]

Conservative Christians drew together across denominational lines to fight for the so-called fundamentals of their traditional faith against the perceived heresy of modernism, and in so doing gave birth to the fundamentalist movement and antievolution crusade. Certainly modernism had made significant inroads within divinity schools and among the clergy of mainline Protestant denominations in the North and West, and fundamentalism represented a legitimate theological effort to counter these advances. Biblical higher criticism and an evolutionary world view, as twin pillars of this opposing creed, stood as logical targets of a conservative counterattack. A purely theological effort, however, rarely incites a mass movement, at least in pluralistic America; much more stirred up fundamentalism—and turned its fury against teaching evolution in public schools.

The First World War played a pivotal role. American intervention, as part of a progressive effort to defeat German militarism and make the world "safe for democracy," was supported by many of the modernists, who revered the nation's wartime leader, Woodrow Wilson, himself a second-generation modernist academic. A passionate champion of peace, William Jennings Bryan opposed this position and in

1915 resigned his post as Wilson's secretary of state in protest over the drift toward war. He spent the next two years criss-crossing the country campaigning against American intervention.

Many leading premillennialists shared Bryan's open hostility toward America's intervention in the European conflict, seeing the war as both a product of the depravity of the age and the possible fulfillment of a prophesy regarding the coming of the next millennium. With Shailer Mathews leading the charge, some modernists used this opportunity to attack premillennialism as an otherworldly threat to national security in wartime. Some premillennialists responded in kind by stressing the German roots of higher criticism, attributing an evolutionary "survival of the fittest" mentality to German militarism and accusing modernism of undermining traditional American faith in biblical values. "The new theology has led Germany into barbarism," the premillennialist journal *Our Hope* declared in a 1918 editorial, "and it will lead any nation into the same demoralization."[9] The trauma of war stirred passions on both sides and helped spur a bitter, decade-long battle among American Christians. "These ideas, and the cultural crisis that bred them, revolutionized fundamentalism," the historian of religion George M. Marsden observed. "Until World War I various components of the movement were present, yet collectively they were not sufficient to constitute a full-fledged 'fundamentalist' movement. The cultural issue suddenly gave the movement a new dimension, as well as a sense of urgency."[10]

When a horribly brutal war led to an unjust and uneasy peace, the rise of international communism, worldwide labor unrest, and an apparent breakdown of traditional values, the cultural crisis worsened for conservative Christians in the United States. "One indication that many premillennialists were shifting their emphasis—away from just evangelizing, praying, and waiting for the end time, toward more intense concern with retarding [social] degenerative trends—was the role they played in the formation of the first explicitly fundamentalist organization," Marsden noted. "In the summer of 1918, under the guidance of William B. Riley, a number of leaders in the Bible school and prophetic conference movement conceived of the idea of the World's Christian Fundamentals Association."[11]

During the preceding two decades, Riley had attracted a 3,000-member congregation to his aging Baptist church in downtown Minneapolis through a distinctive combination of conservative dispensational-premillennialist theology and politicized social activism. "When the Church is regarded as the body of God-fearing, righteous-living men, then, it ought to be in politics, and as a powerful influence," he proclaimed in a 1906 book that urged Christians to promote social justice for the urban poor and workers.[12] During the next decade, Riley focused his social activism on outlawing liquor, which he viewed as a key source of urban problems. By the twenties, he turned against teaching evolution in public schools. Later, he concentrated on attacking communism. Following the First World War and flushed with success upon ratification of the Eighteenth Amendment authorizing Prohibition, he was ideally suited to lead premillennialists into the cultural wars of the twenties.

In 1919, Riley welcomed some 6,000 conservative Christians to the World's Christian Fundamentals Association (WCFA) inaugural conference with the warning that their Protestant denominations were "rapidly coming under the leadership of the new infidelity, known as 'modernism.'" One by one, seventeen prominent ministers from across the country—the future high priests of fundamentalism—took the podium to denounce modernism as, in the words of one speaker, "the product of Satan's lie," and to call for a return to biblical fundamentals in church and culture. "It is ours to stand by our guns," Riley proclaimed in closing the conference. "God forbid that we should fail him in the hour when the battle is heavy."[13] Participants then returned to their separate denominations, ready to battle the modernists. Only minor conflicts erupted within Protestant Episcopal and northern Methodist churches, where modernism was firmly entrenched, or in southern Baptist and Presbyterian congregations, where conservatives encountered little opposition. Both sides proved roughly equal in strength within the northern Baptist and Presbyterian denominations, however, resulting in fierce battles for control. Indeed, it was during the ensuing intradenominational strife within the Northern Baptist Convention that conservative leader Curtis Lee Laws coined the word *fundamentalist* to identify those willing "to do battle royal for the Fundamentals."[14] Use of the term quickly spread to include all conservative Christians militantly opposed to modernism.

Although these early developments laid the foundation for the antievolution crusade and the ensuing Scopes trial, they did not predestine it. Fundamentalism began as a response to theological developments within the Protestant church rather than to political or educational developments within American society. Even the name of the WCFA's journal, *Christian Fundamentals in Schools and Churches*, originally referred to support for teaching biblical fundamentals in divinity schools and churches rather than opposition to teaching evolution in public schools—though it neatly fit the organization's later emphasis. "When the Fundamentals movement was originally formed, it was supposed that our particular foe was the so-called 'higher criticism,'" Riley later recalled, "but in the onward going affairs, we discovered that basic to the many forms of modern infidelity is the philosophy of evolution."[15] Riley was predisposed to make this connection, as suggested by the title to one of his earlier books, *The Finality of the Higher Criticism; or, The Theory of Evolution and False Theology*, but it took William Jennings Bryan to turn the fundamentalist movement into a popular crusade against teaching evolution that led directly to Dayton.

Bryan was not a dispensational premillennialist; he was too optimistic. Certainly he shared with premillennialists a joyful hope in eternal life through faith in Christ. But Bryan did not agree with their view that the Bible prophesied the imminent degeneration of the world in preparation for Christ's second coming. Quite to the contrary, he enjoyed things of this world—particularly politics, oratory, travel, and food—and believed in the power of reform to make life better. Reform took two forms for Bryan: personal reform through individual religious faith and public reform through majoritarian governmental action. He maintained a deep faith in both throughout his life, and each contributed to his final political campaign against teaching evolution. "My father taught me to believe in Democracy as well as Christianity," Bryan observed late in his life.[16] And so the twig was bent, which grew into the tree.

Bryan's crusade against teaching evolution capped a remarkable thirty-five-year-long career in the public eye. He entered Congress in 1890 as a 30-year-old populist Democratic politician committed to roll back the Republican tariff for the dirt farmers of his native Nebraska. His charismatic speaking ability and youthful enthusiasm quickly earned him the nickname The Boy Orator of the Platte. Bryan's greatest speech occurred at the 1896 Democratic National Convention, where he defied his party's conservative incumbent president, Grover

Cleveland, and the eastern establishment that dominated both political parties by demanding an alternative silver-based currency to help debtors cope with the crippling deflation caused by exclusive reliance on limited gold-backed money. Using a potent mix of radical majoritarian arguments and traditional religious oratory, he demanded, "You shall not press down upon the brow of labor this crown of thorns, you shall not crucify mankind upon a cross of gold." The speech electrified the convention and secured the party's presidential nomination for Bryan. For many, he became known as the Great Commoner; for some, the Peerless Leader.

A narrow defeat in the ensuing bitter election did not diminish Bryan's faith in God or the people. He retained leadership of the Democratic party and secured two subsequent presidential nominations as he fought against imperialism and militarism following the Spanish-American War and for increased public control over corporate business practices. His vocation became speaking and writing, with majoritarian political commentary and evangelical Protestant lectures serving as his stock in trade. During the remainder of his life, the energetic Bryan gave an average of more than two hundred speeches each year, traveled continually throughout the country and around the world, wrote dozens of books, and edited a political newspaper with a nationwide circulation. After helping Woodrow Wilson secure the White House in 1912, Bryan became secretary of state and idealistically (some said naively) set about negotiating a series of international treaties designed to avert war by requiring the arbitration of disputes among nations. This became more of a religious mission than a political task for Bryan, who called on America to "exercise Christian forbearance" in the face of increasing German aggression and vowed, "There will be no war while I am Secretary of State."[17] Of course, he had to resign from office to keep this promise.

Once again left without a formal governmental post but with an expanded sense of mission, Bryan resumed his efforts as an itinerant speaker and writer on political and religious topics. Although his campaign for peace failed, he helped to secure ratification of four constitutional amendments designed to promote a more democratic or righteous society: the direct election of senators, a progressive federal income tax, Prohibition, and female suffrage. During this period, the aging Commoner moved to Miami for his wife's health and got in on the ground floor of the historic Florida land boom of the early twenties. Although publicly he played down his profits, the spectacular rise in land prices made Bryan into a millionaire almost overnight.

Private wealth did not diminish Bryan's public zeal as he found two campaign targets: the conservative Republican administrations in Washington and teaching evolution in public schools. Both targets remained fixed in his sights throughout the final years of his life. Indeed, after seeing himself portrayed in a political cartoon as a hunter shifting his aim from a Republican elephant to a Darwinian monkey, Bryan admonished the cartoonist: "You should represent me as using a double-barreled shotgun fixing one barrel at the elephant as he tries to enter the treasury and another at Darwinism—the monkey—as he tries to enter the school room."[18] Bryan remained a progressive even as he crusaded against teaching evolution. "In William Jennings Bryan, reform and reaction lived happily, if somewhat incongruously, side by side," biographer Lawrence W. Levine concluded. "The Bryan of the 1920's was essentially the Bryan of the 1890's: older in years but no less vigorous, no less optimistic, no less certain."[19]

Bryan's antievolutionism was compatible with his progressive politics because both supported reform, appealed to majoritarianism, and sprang from his Christian convictions. Bryan alluded to these issues in his first public address dealing with Darwinism, which he composed in 1904 at the height of his political career. From this earliest point, he described Darwinism as "dangerous" for both religious and social reasons. "I object to the Darwinian theory," Bryan said with respect to the religious implications of a naturalistic explanation for human development, "because I fear we shall lose the consciousness of God's presence in our daily life, if we must accept the theory that through all the ages no spiritual force has touched the life of man and shaped the destiny of nations." Turning to the social consequences of the theory, Bryan added, "But there is another objection. The Darwinian theory represents man as reaching his present perfection by the operation of the law of hate—the merciless law by which the strong crowd out and kill off the weak."[20]

The Great Commoner was no more willing to defer to ivy tower scientists on this issue than to Wall Street bankers on monetary matters. "I have a right to assume," he declared in this early speech, "a Designer back of the design [in nature]—a Creator back of the creation; and no matter how long you draw out the process of creation; so long as God stands back of it you can not shake my faith in Jehovah." This last comment allowed for an extended geologic history and even for limited theistic evolution; but Bryan dug in his heels regarding the supernatural creation of humans and described it as "one of the test questions with the Christian."[21] Although Bryan regularly delivered this speech on the Chautauqua circuit during the early years of the century, he said little else against Darwinism until the twenties, when he began blaming it for the First World War and an apparent decline in religious faith among educated Americans.

As a devout believer in peace, Bryan could scarcely understand how supposedly Christian nations could engage in such a brutal war until two scholarly books attributed it to misguided Darwinian thinking. In *Headquarters Nights*, the renowned Stanford University zoologist Vernon Kellogg, who went to Europe as a peace worker, recounted his conversations with German military leaders. "Natural selection based on violent and fatal competitive struggle is the gospel of the German intellectuals," he reported, and served as their justification "why, for the good of the world, there should be this war."[22] Whereas Kellogg used this evidence to promote his own non-Darwinian view of evolutionary development through mutual aid, Bryan saw it as a reason to suppress Darwinian teaching. The philosopher Benjamin Kidd's *The Science of Power* further explored the link between German militarism and Darwinian thinking by examining Darwin's influence on the German philosopher Friedrich Nietzsche. Bryan regularly referred to both books when speaking and writing against teaching evolution. For example, citing Kidd for his authority, Bryan warned in one of his popular books, "Nietzsche carried Darwinism to its logical conclusion and denied the existence of God, denounced Christianity as the doctrine of the degenerate, and democracy as the refuge of the weakling; he overthrew all standards of morality and eulogized war as necessary to man's development."[23]

A third book had an even greater impact on Bryan and touched an even more sensitive nerve. In 1916, the Bryn Mawr University psychologist James H. Leuba published an extensive survey of religious belief among college students and professors. The result confirmed Bryan's worst fears. "The deepest impression left by these records," Leuba concluded, "is that . . . Christianity, as a system of belief, has utterly broken down." Among students, Leuba reported, "the proportion of disbelievers in immortality increases considerably from the freshman to the senior year in college." Among scientists, he found disbelief higher among biologists than physicists, and higher among scientists of greater than lesser distinction, such that "the smallest percentage of believers is found among the greatest biologists; they count only 16.9 per cent of believers in God."[24] Leuba did not identify teaching evolution as the cause for this rising tide of disbelief among educated Americans, but Bryan did. "Can Christians be indifferent to such statistics?" Bryan asked in one speech. "What shall it profit a man if he shall gain all the learning of the schools and lose his faith in God?"[25] This became his ultimate justification for the Scopes trial.

Parents, students, and pastors soon came forward with stories of their own, which Bryan incorporated into his speeches. "At the University of Wisconsin (so a Methodist preacher told me) a teacher told his class that the Bible was a collection of myths," the Commoner related. "A father (a Congressman) tells me that a daughter on her return from Wellesley told him that nobody believed in the Bible stories now. Another father (a Congressman) tells me of a son whose faith was undermined by [Darwinism] in Divinity School."[26] Bryan's wife later recalled, "His soul arose in righteous indignation when he found from many letters he received from parents all over the country that state schools were being used to undermine the religious faith of their children."[27] Of course, many university professors viewed this as their mission, to the extent that it followed as a byproduct of encouraging critical thought and empirical inquiry in an age of scientific positivism. Stanford University president David Starr Jordan, an eminent evolutionary biologist who later volunteered to aid in the legal defense of John Scopes, spoke for many academics when he dismissed traditional Protestant revivalism as "simply a form of drunkenness no more worthy of respect than the drunkenness that lies in the gutter!"[28]

In 1921, Bryan began speaking widely about the dangers of Darwinian ideas, formulating through repeated articulation before diverse audiences arguments he later used at the Scopes trial. Characteristically, this thrust was marked by a new speech, "The Menace of Darwinism," which Bryan repeatedly delivered during the remaining years of his life and incorporated into a popular new book, *In His Image*. "To destroy the faith of Christians and lay the foundations for the bloodiest war in history would seem enough to condemn Darwinism," Bryan thundered, drawing heavily on evidence from Leuba, Kellogg, and Kidd.[29] A second speech against Darwinism, "The Bible and Its Enemies," joined Bryan's repertoire later that year. The Commoner broke out of the starting blocks so fast that the back cover of the 1921 pamphlet containing the "Enemies" speech already referred to "Mr. Bryan and his crusade against evolution."[30]

In addition to stressing the dangers of Darwinism, both speeches denounced the theory as unscientific and unconvincing. "Science to be truly science is classified knowledge," Bryan maintained, adopting an antiquated definition of science. "Tested by this definition, Darwinism is not science at all; it is guesses strung together."[31] He entertained his audiences with exaggerated accounts of seemingly far-fetched evolutionary explanations for human organs—such as the eye, which supposedly began as a light-sensitive freckle. "The increased heat irritated the skin—so the evolutionists guess, and a nerve came there and out of the nerve came the eye! Can you beat it?" Bryan asked rhetorically. "Is it not easier to believe in a God who can make an eye?"[32] As the historian Ronald L. Numbers noted, "Bryan was far from alone in balking at the evolutionary origin of the eye. Christian apologists had long regarded the intricate design of the eye as 'a cure for atheism,' and Darwin himself had readily conceded his vulnerability on this point."[33] Yet Bryan possessed an uncanny ability to exploit any such weaknesses in his opponent's arguments, at least with respect to winning over a popular audience—the only one that mattered to him. "The scientist cannot compel acceptance of any argument he advances, except as, judged on its merits, it is convincing," the Commoner maintained in defiance of scientific authority. "Man is infinitely more than science; science, as well as the Sabbath, was made for man."[34]

This sort of thinking predisposed Bryan to his later course of seeking a legislative judgment on teaching evolution and accepting a trial by jury to enforce any resulting restriction. Indeed, Bryan's mode of operation and optimistic temperament required offering ready political solutions to outstanding social problems—such as a silver-based currency to promote domestic prosperity or arbitration treaties to secure international peace—and his followers, especially those who called him their Peerless Leader, expected an agenda for action. The Menace of Darwinism speech, however, included only a vague call for "real neutrality" on religious issues in public schools: "If the Bible cannot be defended in those schools it should not be attacked."[35] In the fall of 1921, Bryan gave some meaning to this call by publicly wrangling with University of Wisconsin president Edward Burge over allegedly antireligious teaching at that state institution, but Burge, a distinguished scientist, clearly won the argument when it came to issues of academic freedom for university students and professors. Bryan's speech called on the church to purge itself of modernist and evolutionary influences as well, and Bryan soon sought the top post of the northern Presbyterian church to implement this policy within his own denomination; this involved purely parochial matters, however, for which even Bryan would not seek a governmental remedy. Despite his commanding role, the antievolution crusade lacked a specific political or legal objective for nearly a year.

This situation changed almost overnight. Late in 1921, Kentucky's Baptist State Board of Missions passed a resolution calling for a state law against teaching evolution in public schools. Bryan heard about the resolution in January 1922 and immediately adopted the idea. "The movement will sweep the country, and *we* will drive Darwinism from *our* schools," he wrote to the resolution's sponsor. "We have all the Elijahs on our side. Strength to your arms."[36] Bryan had identified his political objective. Within the month, he was on the spot in Lexington, addressing a joint session of the Kentucky legislature on the proposal. Bryan then spent the next month touring the state in support of such legislation, which lost by a single vote in the state House of Representatives.

The campaign for restrictive legislation spread quickly and all but commandeered the antievolution movement. Fundamentalist leader John Roach Straton began advocating antievolution legislation for his home state of New York in February 1922. J. Frank Norris, pastor of the largest church in the Dallas-Fort Worth area, soon took up the cause in Texas. The evangelist T. T. Martin carried the message throughout the South. By fall 1922, William Bell Riley was offering to debate evolutionists on the issue as he traveled around the nation battling modernism in the church. "The whole country is seething on the evolution question," he reported to Bryan in early 1923.[37] Three years later, these same four ministers became the most prominent church figures to actively support the prosecution of John Scopes.

Riley threw the organizational muscle of the WCFA behind the antievolution crusade hoping to politicize the association by giving it a clear legislative objective. Accordingly, the WCFA jumped to the defense of the Kentucky legislation with an editorial in its spring 1922 newsletter and soon began lobbying for similar bills across the country. Ultimately, its interest in enforcing such legislation helped transform the Scopes trial into a major test of fundamentalist influence in American life.

From its first editorial on the subject, an ominous note sounded from the WCFA. The editorialist (most likely Riley) condemned evolution as scripturally and scientifically unsound, conducive to war, and detrimental to morality. Moreover, the editorial baldly asserted that "great scientists" were divided over the theory of evolution and accused its proponents of attempting to settle the controversy "by imposing the theory upon the rising generation" through the public schools.[38] The conspiracy grew darker the following year when an article by Riley in the WCFA newsletter accused evolutionists of "surreptitiously" sowing their "anarchistic socialistic propaganda."[39] Later, Riley accused teachers of evolution of being atheists who "cannot afford to consent to the creation theory, for that would compel recognition of God."[40] By the thirties, he warned of an "international Jewish-Bolshevik-Darwinist conspiracy" to promote evolutionism in the classroom, and praised Adolph Hitler's effort to foil such conspiracies in Germany.[41] The Ku Klux Klan—an organization Bryan despised—supported antievolution laws for much the same reason, adding Roman Catholics to the list of co-conspirators.

Rather than following Riley in proclaiming a need to combat conspiracies, Bryan propelled the antievolution crusade on majoritarian grounds. Bryan's popular arguments shaped the prosecution's case in Dayton. "Teachers in public schools must teach what the taxpayers desire taught," the Commoner admonished the West Virginia legislature in 1923. "The hand that writes the pay check rules the school."[42] Such reasoning went to the core of Bryan's political philosophy. "The essence of democracy is found in the right of the people to have what they want," he once wrote. "There is more virtue in the people themselves than can be found anywhere else."[43] Bryan consistently espoused this philosophy: from the 1890s, when he commented on one of his election defeats, "The people gave and the people have taken away, blessed be the name of the people," through his campaign for world peace in 1917, when he proposed holding a national referendum before the country went to war, to his antievolution crusade of the 1920s. Indeed, the strength of Bryan's convictions in his fight against teaching evolution sprang from his stated belief that "in this controversy, I have a larger majority on my side than in any previous controversy."[44] He estimated that "nine-tenths of the Christians" in America agreed with his views on evolution.[45] Even though that estimate exaggerated the level of support for antievolution laws, clearly a large number of Americans supported Bryan on the issue, especially in the South.[46] "Have faith in mankind," the Commoner proclaimed, "mankind deserves to be trusted."[47]

Individual rights lost out under this political philosophy. "If it is contended that an instructor has a right to teach anything he likes, I reply that the parents who pay the salary have a right to decide what shall be taught," Bryan maintained.[48] "A scientific soviet is attempting to dictate what is taught in our schools," he warned. "It is the smallest, the most impudent, and the most tyrannical oligarchy that ever attempted to exercise arbitrary power."[49] He gave a similarly facile response to charges that antievolution laws infringed on the rights of nonfundamentalist parents and students. Protestants, Catholics, and Jews shared a creationist viewpoint, Bryan believed, and he sought to enlist all of them into his crusade. As for nontheists, he asserted, "The Christians who want to teach religion in their schools furnish the money for denominational institutions. If atheists want to teach atheism, why do they not build their own schools and employ their own teachers?"[50] Such a position assumed that the separation of church and state precluded teaching the Genesis account in public schools. "We do not ask that teachers paid by taxpayers shall teach the Christian religion to students," Bryan told West Virginia lawmakers, "but we do insist that they shall not, under the guise of either science or philosophy, teach evolution as a fact."[51] He apparently expected them to skip the topic of organic origins altogether, or to teach evolution as a hypothesis.

Bryan's comments reflected the deep ambivalence toward individual rights that underlay his majoritarianism. "No concession can be made to the minority in this country without a surrender of the fundamental principle of popular rule," he once proclaimed with respect to Prohibition.[52] When a conservative Supreme Court began striking down Progressive Era labor laws on the ground that they violated the constitutional rights of property owners, Bryan sought to limit judicial review of legislation. Similarly he could argue about teachers of evolution, "It is no infringement on their freedom of conscience or freedom of speech to say that, while as individuals they are at liberty to think as they please and say what they like, they have no right to demand pay for teaching that which parents and the taxpayer do not want taught."[53] To the extent that American political history reflected a tension between majority rule and minority rights, the Commoner stood for majoritarianism. As Edger Lee Masters observed at the time, to Bryan, "the desideratum was not liberty but popular rule."[54]

In his crusade to rally the people against teaching evolution, Bryan was nearly omnipresent. He gave hundreds of speeches on the topic to audiences across the country, including major addresses to nine different state legislatures in the South and Midwest. Bryan pushed the attack in dozens of popular books and articles, beginning with major pieces in the *New York Times* and *Chicago Tribune.* His syndicated "Weekly Bible Talks," carried in daily newspapers with a combined circulation of over 15 million readers, regularly belabored evolutionists. He personally lobbied countless politicians, school officials, and other public figures on the issue. "Forget, if need be, the highbrows both in the political and college world, and carry this cause to the people," he declared.[55] And the people responded.

Concern about the social and religious implications of Darwinism had been a secondary issue within the church for two generations, and although the rise of fundamentalism revived those concerns for some, it took Bryan to transform them into a major political issue. Even Bryan's wife—his closest confidant, who did not share his enthusiasm on this issue—could not understand the response. "Just why the interest grew, just how he was able to put fresh interest into a question which was popular twenty-five years ago, I do not know," she wrote in 1925. "The vigor and force of the man seemed to compel attention."[56] The view was much the same from Tennessee. "Bryan can provoke a controversy quicker than any other man in public or private life," an editorial in the Memphis *Commercial Appeal* observed three months *before* the Scopes trial. "In a public address about two years ago Mr. Bryan saw fit to take a fling at the Darwinian theory. For several years prior to that day we had heard little about evolution. . . . But the Bryan criticism started another controversy, and evolution has become all but a national issue."[57] Two years earlier, the always hostile *Chicago Tribune* complained, "William Jennings Bryan has half the country debating whether the universe was created in six days."[58]

Bryan wanted more than a heated debate, however; he wanted political reform, and this took time. Most states had part-time legislatures that only met in general session during the first few months of odd-numbered years. Kentucky was an exception, but when its antievolution bill died in 1922, proponents of such legislation had to wait until 1923 for their next shot at lawmaking. The legislatures of six different southern and border states actively considered antievolution proposals during the spring of 1923, with Bryan personally involved in most instances, but only two minor measures passed. Oklahoma added a rider to its public-school textbook law providing "that no copyright shall be purchased, nor textbook adopted that teaches the 'Materialistic Conception of History' (i.e.) the Darwin Theory of Creation versus the Bible Account of Creation."[59] The Florida legislature chimed in with a nonbinding resolution declaring "that it is improper and subversive to the best interest of the people" for public school teachers "to teach as true Darwinism or any other hypothesis that links man in blood relationship to any form of lower life."[60]

The Florida resolution was important because Bryan suggested its language and later claimed that it reflected his "views" on the issue, with one significant exception. "Please note," he explained, "that the objection is not to teaching the evolutionary hypothesis as a hypothesis, but to the teaching of it as true or as a proven fact."[61] Bryan also agreed with the resolution's focus on human evolution. In his Menace of Darwinism speech, he stressed that "our chief concern is in protecting man from the demoralization involved in accepting a brute ancestry," and conceded that "evolution in plant life and animal life *up to the highest form of animal* might, if there were proof of it, be admitted without raising a presumption that would compel us to give a brute origin to man."[62] Bryan asked that Florida legislators outlaw such teaching, however, rather than simply to denounce it as improper; but even on this point, the trusting Commoner added, "I do not think that there should be any penalty attached to the bill. We are not dealing with a criminal class."[63] Cautious legislators in Bryan's adopted state compromised by unanimously passing an advisory resolution rather than a law, thereby avoiding any risk of a lawsuit over their action. Two years later, Tennessee legislators displayed less caution than their Florida counterparts—and less trust in teachers than Bryan—by opting for a criminal law on the subject, including a penalty provision, and applying it to all teaching about human evolution rather than solely to teaching it as true. These changes set the stage for the Scopes trial.

Antievolutionists began targeting Tennessee soon after the 1923 lawmaking season ended without a major victory. Bills to outlaw teaching evolution had died in committees of the Tennessee legislature that year, mostly due to inattention as Bryan and other antievolution leaders campaigned in other states. Now they focused on Tennessee and its neighbor, North Carolina, in anticipation of the 1925 legislative sessions. Bryan gave several antievolution speeches in the two states during this period, including a major address in Nashville on January 24, 1924, attended by most Tennessee state officials. Riley toured the region in 1923, 1924, and 1925, speaking widely in fundamentalist churches and calling on the faithful to drive Darwinism from public schools. The WCFA held major national conferences in both states, further arousing local interest in the topic. Billy Sunday scheduled massive popular crusades in the two states, and encamped in Memphis during the 1925 Tennessee legislative session. T. T. Martin, J. Frank Norris, and John Roach Straton also appeared on several occasions. As a result of these efforts, teaching evolution became a hot political issue in both states during the 1924 elections, with many Democratic candidates vowing to support "Bryan and the Bible."

Tennessee offered a particularly promising target, and one that proved more vulnerable than North Carolina. Memphis, the state's largest city, billed itself as "a Baptist stronghold, the citadel of the denomination," and served as a hub for conservative Protestant publishing.[64] The city's leading daily newspaper, the *Commercial Appeal*, could be counted on to endorse antievolution legislation. The state as a whole was solidly Protestant, with more than 1 million church members—nearly half of them Baptists—out of a total adult population of 1.2 million.[65] Governor Austin Peay, a popular Democratic politician known as The Maker of Modern Tennessee for his progressive reforms, described himself as an "old-fashioned Baptist" and often complained that some of the doctrines taught in public schools undermined religious faith.[66] "Be loyal to your religion," he once advised state college students, "scientists and cranks will seek in vain to better it. The Christian faith of our people is the bedrock of our institutions."[67]

Furthermore, Tennessee had a sufficiently diverse population to raise tensions over recent developments in religion and popular culture. Indeed, Bryan opened his 1924 address in the state capital with the challenge, "I make my religious speech here because Nashville is the center of modernism in the South," presumably referring to the influence of Vanderbilt University and the city's nationally famous progressive cleric, James I. Vance.[68] In addition, racial tensions tore at the seams of Tennessee society and exploded into race riots and Klan violence following the First World War. Antievolutionism promised a return to normalcy.

Local defenders of teaching evolution tried to stem the rising tide. Vance argued for a middle course between fundamentalism and modernism in his books and sermons, all the while pleading for tolerance. Nashville's afternoon newspaper, the *Banner*, regularly denounced antievolutionism and sniped at Bryan's motives. When the Commoner proposed forgiving war debts owned by European nations in exchange for their disarmament, for example, the *Banner* sneered, "Neglecting never an opportunity to secure publicity, out of which he has realized a large private fortune, the distinguished Florida gentleman late of Nebraska, William Jennings Bryan, has evolved one of his picturesque and absurd altruistic ideas."[69] Proposals to outlaw teaching evolution were the real target of this editorial, just as they prompted the *Commercial Appeal* to defend Bryan's "honestly accumulated" wealth and publicize his angry denial, "I am not a millionaire."[70] State and local school officials sought to play down teaching evolution in the hope that the crusade would pass them by, but by 1925 this issue had gained too much momentum in Tennessee to be easily turned aside. A legislative confrontation was inevitable.

"Fundamentalism drew first blood in Tennessee today," a January 20, 1925 article in the *Commercial Appeal* reported, "in the introduction of a bill in the Legislature by Senator [John A.] Shelton of Savannah to make it a felony to teach evolution in the public schools of the state."[71] A day later, John W. Butler offered similar legislation in the House of Representatives.[72] Both legislators had campaigned on the issue and their actions were predictable. Butler justified his proposal on Bryanesque grounds: "If we are to exist as a nation the principles upon which our Government is founded must not be destroyed, which they surely would be if . . . we set the Bible aside as being untrue and put evolution in its place."[73] Butler was a little-known Democratic farmer-legislator and Primitive Baptist lay leader. For him, public schools served to promote citizenship based on biblical concepts of morality. Evolutionary beliefs undermined those concepts. Driven by such reasoning, Butler proposed making it a misdemeanor, punishable by a maximum fine of $500, for a public school teacher "to teach any theory that denies the story of the Divine Creation of man as taught in the Bible, and to teach instead that man had descended from a lower order of animal."[74] Most of Butler's colleagues apparently agreed with this proposal, because six days later the House passed it without any amendments. The vote was seventy-one to five. Although three of the dissenters came from Memphis and one from Nashville, the bill gained the support of both rural and urban representatives, including most delegates from every major city in the state.

The House action reflected overwhelming support for the general concept of limits on teaching evolution rather than any detailed consideration of the pending legislation, which Butler had drafted himself. About the only information on the bill that House members received was a free copy of Bryan's 1924 Nashville speech, which did not offer any specific legislative proposal other than to proclaim that if Christians "cannot teach the views of the majority in the schools supported by taxation, then a few people cannot teach at public expense their scientific interpretation that attacks every vital principle of Christianity."[75] Bryan's Florida antievolution resolution expressly incorporated this seemingly balanced restriction on teaching any theory of origins, but Butler's bill dealt solely with the theory of evolution—and the distinction between the two positions was never discussed by Tennessee legislators.

In fact, for various reasons, no specific features of the proposal received a public airing prior to the House vote. First, the press failed to report the bill's introduction, focusing instead on the earlier Senate proposal. Second, the House education committee recommended passage of the measure without holding a public hearing. At least one committee member did not even know of the committee's action prior to the House vote but after mild protest supported the bill anyway. Finally, the House leadership scheduled the bill for final passage during an afternoon set aside for considering inconsequential and uncontroversial legislation. "Measures were ground out of the hopper with regularity," the *Nashville Banner* reported regarding that afternoon in the House, "and with probably less debate than expressed at a session this far along in the life of the legislature." As the afternoon session proceeded, the House postponed action on any bill arousing prolonged discussion. When the disgruntled education committee member asked to hold over the antievolution bill, Butler objected. "I do not see the need for any further talk," Butler reportedly said, "as everyone understands what evolution means." Another representative supported Butler by calling for an immediate vote, and the measure passed without further comment. At the time, this vote received even less attention within the chamber than the passage of a leash law for "egg-sucking dogs," which at least generated laughter after one member asked how to distinguish which dogs suck eggs.[76]

The speedy House action apparently caught the public off guard. No letters to the editor for or against the Butler bill appeared in any major newspaper of the state prior to the House vote. A deluge of letters followed that action. Further, petitions to the legislature and newspaper editorials on the subject only began appearing after the House passed the bill. Of course, final enactment required action by the Tennessee Senate and governor, so plenty of opportunity remained to influence the outcome. By that time, proponents and opponents had thoroughly rehearsed the arguments that would capture the nation's attention during the Scopes trial.

Opponents of the legislation went to work on the Senate. "It was noticed by me that the 'anti-evolution bill' was passed by the house of representatives yesterday," one letter to the *Nashville Banner* observed. "Is it the intention of all those who advocate free expression . . . to let this pass unprotested?"[77] Clearly not, as scores of opponents wrote letters to state newspapers reflecting the outrage, even shame, undoubtedly felt and expressed by countless Tennesseans over the House action. "Let us not blow out the light as long as the student desires to learn," one writer pleaded.[78] "According to the greatest scientific authorities on earth, evolution is no longer regarded as a theory but an established fact," another declared. "But the legislators persist in hearing the teaching of Billy Bryan, Billy Sunday and all the rest."[79] Comparisons to Galileo's trial and Bruno's execution were commonplace. "I fear we will never stamp out the evolution theory, for old Bruno was burned and old Galileo thrown in prison," a sarcastic writer protested, "and yet the damnable round earth theory is still being taught."[80] The inevitable references to monkeys also appeared. "No one needs better proof of the truthfulness of Darwin's theory than to visit Capitol Hill and view some its occupants," a typical letter joked. "Someone said they sprang from monkeys and that one would be forced to believe they had not sprung very far."[81]

Several state newspapers jumped into the fray at this point. "There is no reason why the discoveries of geology and astronomy should be challenges to Christian faith," an editorial in the *Nashville Tennessean* advised.[82] "The quicker this jackass measure is booted into a waste basket, the better for the cause of enlightenment and progress in Tennessee," the *Rockwood Times* added.[83] The *Chattanooga Times* reprinted an editorial declaring, "Perhaps if there is any other being entitled to share Mr. Bryan's satisfaction at this Tennessee legislature it is the monkey. Surely if the human race is accurately represented by that portion of it in the Tennessee house of representatives, the monkey has a right to rejoice that the human race is no kin to the monkey race."[84]

Tennessee's modernist clerics, although vastly outnumbered by their fundamentalist counterparts, held influential pastorates in several cities and joined in condemning the antievolution bill. Indeed, one liberal preacher gave lawmakers such a tongue-lashing that, after a newspaper reprinted his comments, the House took the unusual step of passing a resolution denying them.[85] Thirteen Nashville ministers, most of them either Presbyterians or Methodists, expressed their opposition in a petition to the Senate.[86] Chattanooga's leading liberal pastor, M. S. Freeman, began a widely publicized series of sermons on modernism by criticizing the proposed statute: "I believe that such laws emanate from a false conception that our Christian faith needs to be sheltered behind bars."[87] The modernist leader R. T. Vann, who played a lead role in opposing antievolution legislation in North Carolina, delivered an address in Memphis on the need for academic freedom in science education. "Now, granted that we may and must teach science in our colleges," he argued, "this teaching must be done by scientists. . . . Neither priest nor prophet nor apostle, nor even our Lord Himself, ever made the slightest contribution to our knowledge of natural science."[88] Already, the three main tactics for attacking the antievolution measure had emerged: the defense of individual freedom, an appeal to scientific authority, and a mocking ridicule of fundamentalists and biblical literalism; later, they became the three prongs of the Scopes defense.

Senators readily responded to these arguments. Just two days after the House passed the Butler bill, the Senate judiciary committee voted down Shelton's antievolution measure with the comment that the legislature should not "make laws that even remotely affect the question of religious belief."[89] Six days later, the committee also rejected the Butler bill. Caught in the middle between its judiciary committee and the House, the full Senate vacillated. First, it considered a move to kill the Butler bill outright, then it adopted a motion to schedule the legislation for an expedited vote. Finally, it sent both antievolution bills back to the judiciary committee for reconsideration during a month-long legislative recess that began in mid-February.[90] This provided time for antievolutionists to counterattack. They made the most of their opportunity.

"As near as we can judge," one reader wrote to the *Nashville Banner* in early February 1925, "the house of representatives has passed the bill . . . , the senate judiciary committee has recommended the bill for rejection . . . , and the forum has thus far been monopolized by those who oppose the bill. It is time something was said for the other side."[91] This writer, and dozens of other antievolutionists who sent letters to the editors of Tennessee newspapers in support of the Butler bill, displayed an understanding and acceptance of Bryan's basic argument that the majority should oversee the content of public school instruction, at least with respect to the teaching of "unproven" theories that profoundly influenced social and spiritual values.

These public letters raised familiar issues. "Why should Christians and other good citizens be taxed to support the groundless guesses of infidels, which are being taught under the pretense of scientific discovery?" one asked.[92] "No one is opposed to research or the word evolution, but ninety-nine per cent of the people of the United States oppose the objectionable teaching that man . . . evolved from some sort of lower animal life," another protested. "If the public is opposed to such teaching it is their inalienable right through the lawmakers to pass a law prohibiting such teaching in schools supported by taxation."[93] Many of these letters were written by women, such as the one asking, "What are mothers to do when unwise education makes boys lose confidence in the home, the Bible, the government and all law?"[94] Such letters expressed the sentiments of many Tennesseans and called for action by the Senate. "I glory in our so-called ignorant Legislature," a self-professed fundamentalist wrote. "Would to God we had more Bryans and fewer Darwin advocates. I do hope that the Senate will concur with the House and pass our evolution act."[95]

On the day that the full Senate voted to revive consideration of the Butler bill, Senator John Shelton sought Bryan's help. "I am writing to know just what form of legislation you would suggest," Shelton inquired. "Other members have asked me to write you for suggestions before the matter comes up for final passage." The Senate sponsor of the legislation then invited Bryan to address a joint session of the legislature following the upcoming recess.[96] Bryan declined this invitation but offered one suggestion on the bill. "The special thing that I want to suggest is that it is better not to have a penalty," he advised. "In the first place, our opponents, not being able to oppose the measure on its merits, are always trying to find something that will divert attention, and the penalty furnishes the excuse. . . . The second reason is that we are dealing with an educated class that is supposed to respect the law."[97] With no penalty, of course, there would be no martyrs to the cause of freedom—and no Scopes trial—simply obstinate schoolteachers flaunting the public will. Bryan could foresee the public relations impact of both courses. On the brink of victory, however, Tennessee crusaders ignored his words of caution.

Two other national fundamentalist leaders did appear on the scene at this time, Billy Sunday and J. Frank Norris—with Sunday having the greater impact. Norris, Riley, and Bryan could preach to the converted and mobilize conservative Protestants into a fighting force; Bryan also could mesmerize a political audience; but no fundamentalist of the twenties could match Sunday's ability to draw a crowd and win converts. A Billy Sunday crusade would hit a town like the arrival of the Ringling Bros. Circus, with Sunday performing in all three rings at once. The former Chicago Cubs outfielder would preach and pray, sing and shout, and leap across the stage delivering rapid-fire sermons before huge audiences.

During February 1925, Sunday broke his custom of spacing his appearances by returning to Memphis for a second crusade in as many years. An opening night audience of more than five thousand heard him proclaim "a star of glory to the Tennessee legislature, or that part of it involved, for its action against that God forsaken gang of evolutionary cutthroats." The crowds grew as the eighteen-day crusade continued, with Sunday regularly denouncing, as he repeatedly described it, "the old bastard theory of evolution." He damned Darwin as an "infidel" on one occasion and shouted, "To Hell with the Modernists," on another, but reserved a special scorn for teachers of evolution. "Education today is chained to the devil's throne," Sunday proclaimed in one typical staccato outburst. "Teaching evolution. Teaching about prehistoric man. No such thing as pre-historic man. . . . Pre-historic man. Pre-historic man," at which point, the *Commercial Appeal* reported, "Mr. Sunday gagged as if about to vomit." Any deeper issues regarding the social and spiritual implications of naturalistic evolution were lost in a superficial plea for biblical literalism. Indeed, one local journalist described Sunday's Memphis crusade as "the most condemnatory, bombastic, ironic and elemental flaying of a principle or a belief that [he] ever heard in his limited lifetime and career from drunken fist fights to the halls of congress."[98]

"All kinds, varieties, and species came out to hear Sunday," wrote the *Commercial Appeal*, which gave daily, front-page coverage to the event. Thousands attended Men's Night, where males could freely show their emotion out of the sight of women. Even more turned out for Ladies' Night. The newspaper reported that "15,000 black and tan and brown and radiant faces glowed with God's glory" on Negro Night. An equal number of "Kluxers"—some wearing their robes and masks—turned out for the unofficial Klan Night. Special trains and buses brought people from all over Tennessee. Many legislators appeared on one or more occasions. Total attendance figures topped 200,000 people, which represented one-tenth of the state's population.[99] By the time Sunday left and Norris arrived, on the eve of the Senate vote, Tennessee fundamentalists were fully aroused. Compared to Sunday's bombast, Norris's suggestion that the evolution teacher "has his hands dripping with innocent blood" sounded downright conciliatory. At least Norris spoke in complete sentences. "Tennessee has before her citizens a bill which aims at the teaching of evolution in the schools," he observed. "I sincerely hope that Tennessee will be the first state to [do so] by enactment of her legislature."[100]

Norris's hope by this time was the people's will and the legislature's intent. On March 11, at its first meeting following the legislative recess, the Senate judiciary committee approved the Butler bill without amendment. Ten days later, the full Senate concurred by a vote of twenty-four to six, and sent the bill directly to Governor Peay for his signature. The scant opposition was scattered among senators from rural and urban districts from both political parties, without any apparent pattern other than personal conviction.

The Senate vote followed a spirited, three-hour floor debate in which proponents stressed majority rule and the religious faith of schoolchildren. Speaker of the Senate L. D. Hill, a devout Campbellite Christian who represented Dayton in the upper house, set the tone. At the outset, he barred consideration of an amendment designed to ridicule the legislation by additionally outlawing instruction in the round-earth theory. He then stepped down from the Speaker's chair to give an impassioned plea for the bill. "I say it is unfair that the children of Tennessee who believe in the Bible literally should be taught things contrary to that belief in the public schools maintained by their parents," the Speaker declared. "If you take these young tender children from their parents by the compulsory school law and teach them this stuff about man originating from some protoplasm or one-cell matter or lower form of life, they will never believe the Bible story of divine creation." Senate antievolution bill sponsor John Shelton added that taxpayers "who believe in the divine creation of man should not be compelled to help support schools in which the theory of evolution is taught." One reluctant supporter justified his vote "on the ground that an overwhelming majority of the people of the state disbelieve in the evolution theory and do not want it taught to their children." A more enthusiastic proponent estimated that this majority included "95 percent of the people of Tennessee."[101]

Outnumbered Senate opponents of the legislation countered with pleas for individual rights. "It isn't a question of whether you believe in the Book of Genesis, but whether you think the church and state should be kept separate," one senator asserted. "No law can shackle human thought," another declared. A Republican lawmaker quoted passages on religious freedom from the state constitution, and blamed the entire controversy on "that greatest of all disturbers of the political and public life from the last twenty-eight or thirty years, I mean William Jennings Bryan." But a proponent countered, "This bill does not attempt to interfere with religious freedom or dictate the beliefs of any man, for it simply endeavors to carry out the wishes of the great majority of the people." Such sentiments easily carried the Senate.[102]

State and national opponents of antievolution laws appealed to Governor Peay to veto the legislation. Owing to the governor's national reputation as a progressive who championed increased support for public education and a longer school year—efforts that later led to the naming of a college in his honor—those writing from out of state probably entertained some hope for success. Urged on by the California science writer Maynard Shipley and his Science League of America, a new organization formed to oppose antievolutionism, letters of protest poured in from across America. For example, taking the line of Draper and White, a New Yorker asked, "The Middle Ages gave us heretics, witches burnt at the stake, filth and ignorance. Do we want to return to the same?" From within Tennessee, some concerned citizens appealed for a veto. The dean of the state's premiere African-American college, Fisk University, wrote, "As a clergyman and educator, I hope that you will refuse to give your support to the Evolution Bill. It would seem most unfortunate to me should the State of Tennessee legislate against the beliefs of liberal Christianity." The Episcopal bishop of Tennessee added, "I consider such restrictive legislation not only unfortunate but calamitous."[103]

Yet most letters to the governor from Tennesseans supported the measure, and two potentially significant opponents kept silent. The University of Tennessee's powerful president Harcourt A. Morgan, who privately opposed the antievolution bill, held his tongue so long as Peay's proposal for expanding the university still awaited action in the state legislature—and admonished his faculty to do likewise. In a confidential note, he assured the governor, "The subject of Evolution so intricately involves religious belief, which the University has no disposition to dictate, that the University declines to engage in the controversy." Only after the legislature adjourned and the new law became the primary subject of ridicule at the annual student parade did the depth of university opposition to it become apparent.[104] Similarly, the Tennessee Academy of Sciences, which counted Morgan and other university scientists among its leading members, said nothing against the measure until after it became law. This left Peay free to follow his personal and political inclinations. "I remember a short conversation I had with you on the capitol steps some weeks ago about the Evolution bill. You said then, 'That you thought you would sign it,' " a Nashville minister wrote to the governor. "May I, as your friend and supporter, ask you to sign the present bill and help us in Tennessee who are making a desperate fight against the inroads of Materialism."[105] Peay kept his word.

The governor explained his decision to sign the bill in a curious message to the legislature. On one hand, Peay firmly asserted for proponents, "It is the belief of our people and they say in this bill that any theory of man's descent from lower animals, . . . because a denial of the Bible, shall not be taught in our public schools." On the other hand, he assured opponents that this law "will not put our teachers in any jeopardy." Indeed, even though the most cursory review of Tennessee high school biology textbooks should have shown him otherwise, Peay wrote, "I can find nothing of consequence in the books now being taught in our schools with which this bill will interfere in the slightest manner." Nevertheless, he went on to hail the measure as "a distinct protest against an irreligious tendency to exalt *so-called* science, and deny the Bible in some schools and quarters—a tendency fundamentally wrong and fatally mischievous in its effects on our children, our institutions and our country."[106]

Peay, whose progressivism grew out of his traditional religious beliefs, simply could not accept a conflict between public education and popular religion. In 1925, only days after he signed the antievolution law, Peay won legislative approval for a massive education reform bill that laid the foundation for Tennessee's modern, state-supported system of public schools. He approved of limits on teaching evolution as part of those state-funded schools. "I have profound contempt for those who are throwing slurs at Tennessee for having the [antievolution] law," Peay said during the Scopes controversy. "In my judgement any state had better dispense with its schools than with its Bible. We are keeping both." Yet he could not totally ignore the tension between a fundamentalist's fear of modern education and a progressive's faith in it. In his message to the legislature on the antievolution bill, he fell back on Bryan's populist refrain: "The people have a right and must have the right to regulate what is taught in their schools."[107] Trapped between fundamentalism and progressivism, Peay may have viewed majoritarianism as an excuse for the law. Caught in the same bind, Bryan saw it as the law's ultimate justification.

Bryan rejoiced in the decision by Peay to sign the legislation, but both individuals misjudged the consequences of that action. "The Christian parents of the State owe you a debt of gratitude for saving their children from the poisonous influence of an unproven hypothesis," Bryan wired the governor. "Other states North and South will follow the example of Tennessee."[108] He missed the mark with this prediction because he failed to anticipate a test case involving the act. Indeed, blinded by their sunny progressive faith in the curative power of majoritarian reforms, neither of these experienced politicians saw the Scopes trial coming. According to Peay, the law simply covered anti-biblical doctrines, and he trusted that "nothing of that sort is taught in any accepted book on science." Bryan, for his part, trusted that public school teachers would "respect the law." Both could agree with Peay's comment, "Nobody believes that it is going to be an active statute."[109] It took Riley and the WCFA to appreciate the potential significance of an incipient conspiracy by free-speech advocates, evolution scientists, and publicity-minded townspeople for testing the law in court—and then to call on Bryan to defend his majoritarian reform against charges that it violated elemental concepts of individual liberty.

— CHAPTER THREE —

In Defense of Individual Liberty

Activists with the American Civil Liberties Union did not dismiss the enactment of the Tennessee law against teaching evolution as an insignificant occurrence in some remote intellectual backwater. More critically, they did not view the antievolution crusade in isolation; if they had, they probably would have ignored it along with countless other laws and movements to advance Protestant culture then prevalent throughout the United States. Prior to the Scopes trial, the ACLU did not display any particular interest in challenging government efforts to protect or promote religious beliefs. To the contrary, Quakers played a major role in founding and financing the organization during the First World War as a vehicle to protect religiously motivated pacifists from compulsory military service. Yet ACLU leaders saw the new Tennessee statute in a different light, one that made it stand out as a threat to freedom and individual liberty in the broader American society.

A fashionable new book of the era, *The Mind in the Making* by James Harvey Robinson of the left-wing New School for Social Research in New York City, captured the reactionary mood of the times as perceived by many of the socially prominent, politically radical New Yorkers who led the ACLU during the early twenties. According to this book, which incorporated an evolutionary view of intellectual and social history, a systematic assault on personal liberty in the United States began during the First World War; various state and local authorities had limited freedom prior to this period, to be sure, but these earlier restrictions represented isolated incidents and could be dealt with accordingly. The war changed everything.[1]

"It is a terrible thing to lead this great and peaceful people into war," President Wilson declared in his 1917 war message to Congress. He then added to the terror of some by warning that "a firm hand of stern repression" would curtail domestic disloyalty during wartime.[2] At Wilson's request, Congress imposed a military draft, enacted an Espionage Act that outlawed both obstructing the recruitment of troops and causing military insubordination, and authorized the immigration service to denaturalize and deport foreign-born radicals. The federal Justice Department broadly construed the Espionage Act to cover statements critical of the war effort, while the postal service revoked mailing privileges for publications it considered to "embarrass or hamper the government in conducting the war."[3] In 1918, Congress responded to mounting domestic opposition to the war by expanding the Espionage Act to bar "disloyal" or "abusive" statements about the American form of government. Several states outlawed teaching German in public schools. Public and private institutions of higher education throughout the United States, including Columbia University, in the ACLU's backyard, dismissed tenured faculty members for opposing American intervention in the war.

The National Civil Liberties Bureau was established in 1917 to defend conscientious objectors and antiwar protesters. It initially grew out of the American Union Against Militarism (AUAM), an organization formed by wealthy pacifists to oppose American entry into World War I, but it soon acquired a separate existence under the leadership of Roger Baldwin, a Harvard-educated social worker with an aristocratic pedigree and radical leanings. "We are interested in preserving civil liberties in America," Baldwin explained at the time, "first, for the sake of democracy itself, and second, for the rights of the people to discuss peace terms and war policies. The rights of both individuals and minorities are being grossly violated throughout the country."[4]

Such violations struck close to home for the fledgling organization and helped shape its libertarian philosophy toward free speech. Within weeks of the bureau's formation, the postal service banned from the mail twelve different antiwar pamphlets the bureau had prepared for mass distribution. Anticipating this problem, Baldwin sought prepublication approval for mailing the pamphlets and secured the aid of Clarence Darrow to negotiate a settlement with the Postmaster General, but to no avail. One of the banned pamphlets, authored by bureau officer and future American Socialist leader Norman M. Thomas, articulated the view of democracy and liberty then taking hold among bureau activists. President Wilson, who hailed from the same liberal establishment that gave birth to the bureau, maintained a majoritarian view of democracy that justified restrictions on free speech and other minority rights once Congress declared war. In the bureau pamphlet, Thomas countered that the majority should never assert control over matters of individual conscience. "Democracy degenerates into mobocracy unless the rights of minorities are respected," Thomas wrote.[5] Triggered both by this encounter with the postal service and by the bureau's subsequent defense of the radical Industrial Workers of the World (IWW), federal agents began spying on bureau activities. In 1918, Justice Department officials raided the bureau's headquarters and threatened to indict organization leaders. Later that year, Baldwin began a year in prison for refusing to comply with the Selective Service Act.

Bureau leaders initially tried to cooperate with officials in the Wilson administration, many of whom also came from elite backgrounds. Indeed, prior to 1918, Baldwin naively supplied sensitive bureau information to his friends in high government positions, confident that this would ease official suspicions about the organization and its supporters; instead, it led to mass arrests. At the time of his own imprisonment, Baldwin could only comfort his mother by writing that "my guardian [is] a fine young Yale man."[6]

These bitter experiences gradually changed the outlook toward democratic government held by Baldwin and other members of the bureau's executive committee. "Largely oblivious to civil liberties considerations before the war, the wartime crisis forced them to abandon their faith in the inevitability of social progress and their majoritarian view of democracy," the ACLU historian Samuel Walker concluded. "They now began to see that majority rule and liberty were not necessarily synonymous and thus discovered the First Amendment as a new principle for advancing human freedom."[7] This new antimajoritarian impulse, forged in the crucible of wartime mass hysteria, profoundly influenced the ACLU's response to the antievolution crusade.

Proponents of civil liberties expected conditions to improve after the armistice in 1918, but to them the repression appeared only to intensify. "The war brought with it a burst of unwanted and varied animation. . . . It was common talk that when the foe, whose criminal lust for power had precipitated the mighty tragedy, should be vanquished, things would 'no longer be the same,'" Robinson wrote. "Never did bitter disappointment follow such high hopes. All the old habits of nationalistic policy reasserted themselves at Versailles. . . . Then there emerged from the autocracy of the Tsars the dictatorship of the proletariat, and in Hungary and Germany various startling attempts to revolutionize hastily and excessively." From these developments the so-called Red Scare ensued. "War had naturally produced its machinery for dealing with dissenters, . . . and it was the easiest thing in the world to extend the repression to those who held exceptional or unpopular views, like the Socialists and members of the I.W.W.," Robinson reasoned. "But suspicion went further so as to embrace members of a rather small, thoughtful class who, while rarely socialistic, were confessedly skeptical in regard to the general beneficence of existing institutions, and who failed to applaud at just the right points to suit the tastes of the majority of their fellow-citizens."[8] Robinson identified with this latter class, as did many bureau activists and their friends. As a progressive reformer prior to the war, for example, Baldwin campaigned for many majoritarian reforms, such as the initiative and referendum process, but increasingly he shifted his zeal to the defense of individual rights as he suffered under the excesses of majority rule.

Events drove the Red Scare. An unprecedented number of strikes paralyzed large sectors of American business during 1919 as an epidemic of labor unrest swept the country following the war. A general strike in Seattle and a police strike in Boston threatened public safety. Race riots broke out in several cities—including Knoxville, just down the road from the future site of the Scopes trial. Terrorist bombings rocked the home of U.S. Attorney General A. Mitchell Palmer and mail bombs were sent to dozens of other political and business leaders. Newly formed domestic Communist parties defended violent revolution abroad and labor militancy at home, the two seeming to blur in the minds of many frightened Americans. "The circumstances of our participation in the World War and the rise of Bolshevism convinced many for the first time that at last society and the Republic were actually threatened," Robinson observed.[9]

The government reacted swiftly. Most states outlawed the possession or display of either the red flag of communism or the black flag of anarchism. They also enacted and strictly enforced tough new "criminal syndicalism" laws against organized violent or unlawful activities designed to disrupt commercial or governmental activities. In the bureau's home state, the legislature formed a special panel known as the Lusk Committee to combat revolutionary radicalism. This committee's massive report relied partially on confiscated bureau files to expose, as the report described them, "various forces now at work in the United States, and particularly within the state of New York, which are seeking to undermine and destroy, not only the government under which we live, but also the very structure of American society."[10] The Lusk Committee swept with a broad broom. Its report condemned socialism, communism, anarchism, bolshevism, pacifism, the international labor movement, and, of course, the bureau, which was called "a supporter of all subversive groups."[11] Its chief counsel arrested hundreds of New Yorkers associated with these movements.

The Wilson administration in Washington supplemented such state actions with a series of coordinated police raids that ransacked the homes and offices of alleged radicals across the country and led to the arrest and prolonged detention of thousands of suspects, often without valid warrants or court orders. Late in 1919, the Justice Department deported to the Soviet Union a shipload of denaturalized Communists. Radical labor leaders bore the brunt of these assaults. Yet the liberal Democratic administration of Woodrow Wilson did not go far enough for many Americans. Republicans recaptured the White House in 1920 with a presidential candidate who promised a "return to normalcy" and a vice presidential candidate who had broken the Boston police strike and championed the cause of immigration restriction. Civil liberties remained in jeopardy.

"Well, of course, it was a time of tremendous labor unrest, highlighted by the two general strikes in the steel mills and coal mines. And it was also, and I guess above all, a time of intense radical agitation, brought on by the Russian Revolution," Roger Baldwin later recalled. "So by the time the World War was over we had a new war on our hands—a different one. Then, instead of arresting and persecuting opponents of the war, we were arresting and persecuting friends of Russia."[12] Thus events stood when Baldwin left prison and reassumed leadership of the National Civil Liberties Bureau. He promptly concluded, as he stated in a memorandum to the executive committee, that the bureau should be "reorganized and enlarged to cope more adequately with the invasions of civil liberties incident to the industrial struggle which had followed the war." Direct action to protect labor unions would replace legal maneuvers on behalf of pacifists as the bureau's principal focus. The bureau assumed a new name to go with its new mission: the American Civil Liberties Union. "The cause we now serve is labor," Baldwin proclaimed at the time, and labor included public school teachers.[13]

The new cause and methods adopted by the ACLU set the stage for how it would handle the Scopes trial. It remained an elitist organization dominated by liberal, educated New Yorkers who had grown wary of majoritarianism. Instinctively they opposed popular movements to restrict academic freedom, such as the antievolution crusade, but failure to achieve judicial redress for their grievances, especially on behalf of labor unions, led them increasingly to resort to direct action tactics designed to enlighten public opinion. Litigation in and of itself did not hold much promise for protecting minority rights.

The bureau enjoyed some success in providing legal counsel to conscientious objectors during the war, but it failed to make any headway in court toward protecting freedom of expression for antiwar protestors. The ACLU fared no better in its initial courtroom efforts to defend labor organizers following the war. In fact, at the time of the Scopes trial in 1925, the ACLU was still looking for its first court victory. From a legal standpoint, the problem was twofold: states and municipalities imposed many of the objectionable restrictions on speech and assembly, particularly against labor unions, but First Amendment guarantees for freedom of speech, press, assembly, and religion only applied to restrictions by the *federal* government. The Fourteenth Amendment, however, forbade states from depriving "any person of life, liberty, or property, without due process of law." Supreme Court justice John M. Harlan had long maintained that the "liberty" protected against state action by the

Fourteenth Amendment incorporated the basic freedoms enumerated in the First Amendment and other provisions within the Bill of Rights. The full Court did not begin to adopt this position until 1925. That year, in the ACLU-handled appeal of New York Communist leader Benjamin Gitlow's conviction under state law, it selectively incorporated the First Amendment freedoms of speech and press into, as the Court wrote, "the fundamental personal freedoms and 'liberties' protected by the due process clause of the Fourteenth Amendment from impairment by the states."[14] This decision occurred too late and in a too-limited fashion to bolster the ACLU's legal case against the Tennessee antievolution statute in the Scopes trial. Indeed, this potentially momentous development in constitutional jurisprudence did not even help Gitlow, who still lost the case.

Gitlow's defeat highlighted the second legal barrier obstructing the ACLU's efforts to secure free speech rights for antiwar protesters and labor organizers. Federal courts gave little meaning to the First Amendment. The first constitutional challenge to the federal Espionage Act reached the U.S. Supreme Court in 1919, when a unanimous bench upheld the conviction of the Socialist leader Charles T. Schenck for encouraging draft-age men to resist conscription. On the extent of constitutional protection for political speech, the great progressive jurist Oliver Wendell Holmes wrote for the Court, "The question in every case is whether the words used are used in such circumstances and are of such a nature as to create a clear and present danger that they will bring about the substantive evil that Congress has a right to prevent." Congress had a right to protect recruitment and conscription of troops during wartime, the aged Civil War veteran reasoned, and because Schenck's words had a "tendency" to frustrate that effort, the government could stop them.[15]

This sort of reasoning offered scant protection to speech because—as Holmes acknowledged in a letter to New York federal judge Learned Hand, who had close ties to the ACLU—free speech "stands no differently than freedom from vaccination," a freedom that the majority could freely override for the general good.[16] Hand and the ACLU vehemently argued that free speech merited special protection from the majority owing to its unique role in a democracy. Holmes came around to this position in another 1919 Espionage Act decision, *Abrams v. United States*, in which he proposed supplementing his "clear and present danger" test with the qualification, "It is only the present danger of *immediate* evil . . . that warrants Congress in setting a limit to the expression of opinion where private rights are not concerned." According to Holmes's revised view, the "free trade of ideas" in a democracy required protection for political speech unless "an immediate check is required to save the country."[17] Holmes now wrote in dissent, however. A majority of the Court clung to the old view of the First Amendment.

The prevailing judicial interpretation of the First and Fourteenth Amendments offered little prospect that the ACLU could protect free speech through the courts; therefore it adopted other means. "By demonstrations, publicity, pamphlets, legal aid, bail, test cases in courts, financial appeals—by all these methods of daily service the friends of progress to a new social order make common cause," the ACLU's first annual report declared. "The chief activity necessary is publicity in one form or other, for ours is a work of propaganda—getting facts across from our point-of-view."[18] The ACLU ended up fighting many of its battles in court solely because that was where the government took those whom the ACLU sought to defend. The first instinct of the ACLU's founders was to join labor organizers on the picket lines and at mass meetings. Before reassuming leadership of the ACLU, for example, Baldwin spent three months as a laborer in a series of different working-class jobs as a means to study labor conditions firsthand.

The legal community played a surprisingly small role in founding the ACLU. Only three attorneys served on the organization's initial executive committee, and all three supported direct action (rather than litigation) in the fight for civil liberties. During the early 1920s, ACLU representatives spoke at union meetings, organized labor demonstrations, investigated efforts to break strikes, published reports on the plight of workers, and sought legislation to limit antilabor court injunctions and end wartime restrictions on free speech. Baldwin, who was not a lawyer, went so far as to maintain that courts would *never* guarantee civil liberties because rights "are not granted" by those in power.[19] Political radicals and civil rights leaders generally shared this perspective; no public interest law firms existed at the time, and the other early civil rights groups—the NAACP, the Anti-Defamation League, and the American Jewish Congress—mostly relied on publicity and organization to advance their causes. This view of civil liberties litigation—that at most it could publicize an injustice—would shape the ACLU's legal strategy in the Scopes trial.

The most influential lawyer on the ACLU executive committee at the time of the Scopes trial, Arthur Garfield Hays, personified the direct action approach to the fight for civil liberties. A left-wing Park Avenue attorney named by his Republican father after a string of conservative presidents, Hays grew rich and bored representing major corporations and famous entertainers. ACLU activities served as his major diversion for three decades. As Hays wrote in his autobiography, these activities "brought me in contact with a variety of circles, usually poor, defenseless, and unpopular, always the dissenter and persecuted." Championing their right to be heard, Hays advocated an absolutist position on free speech that opposed all government restrictions on "the expression of opinion of any kind, at any time, by anyone or anywhere."[20] This became his mission. "To-day you can talk on any subject you please," Hays wrote in the twenties, "except on a subject which, as a burning issue, would most profit by untrammeled discussion. Speech and assembly are free in New Jersey, West Virginia and Pennsylvania, except to union men in time of strike. If you talk labor unionism then, you land in jail. I know it because I've tried it and I landed in jail."[21]

Hays's personal commitment to direct action on behalf of free speech made him a key actor in many of the ACLU's legendary exploits during the twenties and thirties. He peddled banned books with the writer H. L. Mencken on the Boston Commons in public defiance of a censorship law. Despite a threat that "they'll tar and feather you and castrate you," he confronted mine owners in a strike-bound West Virginia coal town following the murder of three union officials. He defied a ban on public meetings by the strike-busting mayor of Jersey City by delivering an impassioned plea for free speech from atop a car. These experiences made Hays deeply distrustful of majoritarianism and contemptuous of the courts. "We should bear in mind the fact that there may be no greater oppression than by rule of majority," Hays observed at the time. "Tyranny no less exists when imposed by part of a written constitution."[22] In one extreme example of politicized litigation and an act of personal courage for someone of Jewish ancestry, he once ventured into Nazi Germany to defend radicals accused of burning the Reichstag. "Hays was cynical about the legal process and saw court proceedings as a platform for broad and philosophical statements, an opportunity to educate both the judge and the public," Walker observed. "He was simultaneously idealistic about the Bill of Rights and cynical about the courts."[23] Significantly, Hays served as chief ACLU counsel at the Scopes trial.

The ACLU helped set the stage for a show trial in Dayton not only with confrontational methods of promoting free speech but also by its commitment to defend the rights of organized labor. This tie to labor kept the ACLU in close contact with the nation's premier legal defender of radical labor leaders, Clarence Darrow. "I owe the Union more than variety and excitement, more than tang and the 'salt' of life," Hays later recalled. "There began my association with Clarence Darrow. Nothing in life do I treasure more than that, nothing has been more inspiring or humanly helpful than his company, his example, and his friendship."[24]

By the twenties, Darrow unquestionably stood out as the most famous—some would say infamous—trial lawyer in America. Born into an educated, working-class family in rural Ohio, Darrow first gained public notice in the 1890s as a Chicago city attorney and popular speaker for liberal causes. He secured the Democratic nomination to Congress in 1896, but spent most of his time campaigning for the party ticket, headed by presidential nominee William Jennings Bryan, and lost by about one hundred votes. Darrow took up the cause of labor about this time, beginning with the defense of the famed Socialist labor leader Eugene V. Debs against criminal charges growing out of the 1894 Pullman strike. "For the next fifteen years Clarence Darrow was the country's outstanding defender of labor, at a time when labor was more militant and idealistic and employers more hardened and desperate than ever before or since," the liberal *Nation* observed during the Scopes litigation. "The cases he was called upon to defend were almost invariably criminal prosecutions in bitterly hostile communities."[25] The final such case, a dramatic 1911 murder trial involving two union leaders accused of blowing up the *Los Angeles Times* building, tarnished Darrow's reputation with labor when the defendants confessed their guilt after Darrow had professed their innocence.

Thereafter, Darrow gradually shifted his practice to criminal law, defending an odd mix of political radicals and wealthy murderers. Both types of cases kept Darrow's name in the national headlines. The former type also connected the Chicago attorney with the New York-

Clarence Darrow at the time of the Scopes trial. (Courtesy of Bryan College Archives)

based ACLU: they joined forces to defend Benjamin Gitlow, for example. The latter type generated the most publicity, such as the 1924 Leopold–Loeb case, one of the most sensational trials in American history, in which Darrow used arguments of psychological determinism to save two wealthy and intelligent Chicago teenagers from execution for their cold-blooded murder of an unpopular former schoolmate, a crime that the defendants apparently committed for no other reason than to see if they could get away with it. Although Darrow's defense outraged many Americans who believed in individual responsibility, it reflected his long-standing and oft-proclaimed repudiation of free will. [26]

Darrow was not content with simply questioning popular notions of criminal responsibility, but delighted in challenging traditional concepts of morality and religion. One historian described Darrow as "the last of the 'village atheists' on a national scale," and in this role he performed for America the same part that his father once played in his hometown.[27] "He rebelled, just as his father had rebelled, against the narrow preachments of 'do gooders,'" Darrow biographer Kevin Tierney concluded. "He regarded Christianity as a 'slave religion,' encouraging acquiescence in injustice, a willingness to make do with the mediocre, and complacency in the face of the intolerable."[28] In the courtroom, on the Chautauqua circuit, in public debates and lectures, and through dozens of popular books and articles, Darrow spent a lifetime ridiculing traditional Christian beliefs. He called himself an agnostic, but in fact he was effectively an atheist. In this he imitated his intellectual mentor, the nineteenth-century American social critic Robert G. Ingersoll, who wrote, "The Agnostic does not simply say, 'I do not know [if God exists].' He goes another step, and he says, with great emphasis, that you do not know. . . . He is not satisfied with saying that you do not know—he demonstrates that you do not know, and he drives you from the field of fact."[29]

Good intentions underlay Darrow's efforts to undermine popular religious faith. He sincerely believed that the biblical concept of original sin for all and salvation for some through divine grace was, as he described it, "a very dangerous doctrine"—"silly, impossible and wicked."[30] Darrow once told a group of convicts, "It is not the bad people I fear so much as the good people. When a person is sure that he is good, he is nearly hopeless; he gets cruel—he believes in punishment."[31] During a public debate on religion, he added, "The origin of what we call civilization is not due to religion but to skepticism. . . . The modern world is the child of doubt and inquiry, as the ancient world was the child of fear and faith."[32]

Darrow often invoked the idea of organic evolution to support his arguments, but it was never central to his thinking. He claimed to understand modern biology but mixed up Darwinian, Lamarckian, and mutation-theory concepts in his arguments, utilizing whichever best served his immediate rhetorical purposes. He frequently appealed to science as an objective arbitrator of truth, but would only present scientific evidence that supported his position. In short, he was a lawyer. In public debates on religious topics, for example, when confronted with the popular defenses of theism offered by such leading scientists as Arthur Eddington, James Jeans, and Robert Millikan, Darrow would dismiss their expertise as not involving religion and their evidence as hearsay. In contrast, Darrow readily embraced the antitheistic implications of Darwinism.[33]

Darrow's social views shaped his scientific ideas rather that the other way around, and the theory of evolution proved most helpful in his efforts to debunk the biblical notions of creation, design, and purpose in nature. "From where does man get the [selfish] idea of his importance? He gets it from Genesis, of course," Darrow wrote in his autobiography. "In fact, man never was made. He was evolved from the lowest form of life." This view, Darrow maintained, provided a better basis for morality than traditional Christian concepts of eternal salvation and damnation by observing, "No one can feel this universal [evolutionary] relationship without being gentler, kindlier, and more humane toward all the infinite forms of beings that live with us, and must die with us."[34] Of course, Darrow could present a different face in court, such as during the Leopold–Loeb case, when he sought mercy for the defendants by attributing their actions to misguided Social Darwinist thinking.

Darrow welcomed the hullabaloo surrounding the antievolution crusade. It rekindled interest in his legalistic attacks on the Bible, which once appeared hopelessly out of date in light of modern developments in mainline Christian thought. In response to 1923 comments about evolution by William Jennings Bryan, for example, Darrow could again make front-page headlines in the *Chicago Tribune* by simply asking Bryan questions such as, "Did Noah build the ark?" and if so, "how did Noah gather [animals] from all the continents?"[35] Leading Chicago ministers complained that both Bryan's comments and Darrow's questions missed the point, but the public loved it.[36] So did Darrow. When the Scopes trial arose two years later, Darrow volunteered his service for the defense—the only time he ever offered free legal aid—seeing a chance to grab the limelight and debunk Christianity. "My object," Darrow later wrote, "was to focus the attention of the country on the programme of Mr. Bryan and the other fundamentalists in America."[37]

Neither Scopes in particular nor free speech in general mattered much to Darrow, and this troubled many within the ACLU leadership. Baldwin wanted the focus on academic freedom. Acting chair John Haynes Holmes, a liberal Unitarian minister, later complained that Darrow "in the thought processes [regarding religion] was a mid-Victorian arrived too late on the scene."[38] During the twenties, the ACLU executive committee was never openly hostile toward religion per se, and several of its members feared that Darrow's militant agnosticism would imperil Scopes's defense. In his autobiography, Hays clearly took the ACLU's opposition to antievolution laws out of an antireligious context when he wrote, "We have insisted upon the propagandizing rights of various groups—Communists, IWW's, evolutionists, birth-controllers, union organizers, industrialists, freethinkers, Jehovah's Witnesses, and even of Fascists, Nazis, and Lindbergh."[39] Yet in Darrow he saw a soulmate who could turn a small Tennessee courtroom into an international grandstand. Together they went to Dayton.

The ACLU's commitment to civil liberties for workers did more to influence the course of the Scopes trial than simply bringing together Darrow and Hays; it also fired the organization's opposition to antievolution laws. Henry R. Linville, president of the Teachers Union of New York City, served on the ACLU executive committee throughout the twenties. Linville held a doctorate in zoology from Harvard University and chaired the biology department at New York's DeWitt Clinton High School when that institution developed the modern secondary school biology curriculum. George W. Hunter, author of the book at issue in the Scopes trial, had been Linville's colleague at Clinton and his successor as chair of the school's biology department. Linville's own high school texts stressed evolutionary concepts and presented humans within the context of their biological environment.[40] Linville brought to the ACLU both a biology teacher's firsthand experience with instruction in evolution and a labor leader's commitment to protecting free speech and academic freedom for all public school teachers.

Academic freedom had been an ongoing concern of the ACLU from the organization's inception; naturally, it related to free speech, yet the interest ran even deeper. The pacifists who helped form the National Civil Liberties Bureau abhorred wartime efforts to promote patriotism and militarism in the schools. They defended teachers fired for opposing American involvement in the war and fought against efforts to purge the public school curriculum of German influences. After the war, when the ACLU turned its attention to defending unpopular speakers, its efforts widened to include fighting classroom restrictions on unpopular ideas. "The attempts to maintain a uniform orthodox opinion among teachers should be opposed," the ACLU's initial position statement declared. "The attempts of education authorities to inject into public schools and colleges instruction propaganda in the interest of any particular theory of society to the exclusion of others should be opposed."[41]

This statement primarily reflected the ACLU's opposition to school patriotism programs. Building on wartime developments in New York, the Lusk Committee proposed legislation in 1920 to dismiss public school teachers who "advocated, either by word of mouth or in writing, a form of government other than the government of the United States."[42] The ACLU helped persuade New York governor Al Smith to veto this bill in 1921, but Smith's successor signed similar legislation into law a year later. Dozens of other states required public school teachers and college professors to sign loyalty oaths. Powerful patriotic organizations, including the American Legion, lobbied for promoting "Americanism" in the public schools by mandatory patriotic exercises (typically a flag salute) and through classroom use of education materials that praised the military and disparaged all things "foreign" (often including the international labor movement). Publicity generated by the ACLU forestalled these programs in some places, but an ACLU lawsuit challenging compulsory military training for male students attending the state University of California at Los Angeles failed. The rise of a militantly anti-Catholic Ku Klux Klan during the early 1920s led to ACLU efforts to protect both Catholic teachers from mass firings in Klan-dominated school districts and the free-speech rights of the Klan in Catholic communities. Repeatedly, the ACLU was drawn into courtrooms over education. Indeed, during the 1920s, it had to go to court to protect its own right to sponsor programs in New York City schools after the local board of education barred all ACLU representatives from "talking in school buildings" under a general regulation requiring classroom speakers to "be loyal to American institutions."[43]

Attempts to propagandize public education did not begin in the twenties. In fact, Massachusetts Puritans founded America's first public schools during the colonial era partly to promote their distinctive religious and political system. The common-school movement spread during the nineteenth century (at least in part) as a means to indoctrinate into American ways the large number of non-English immigrants entering the United States. Most public school curricula traditionally included American civics, Bible reading, and daily prayers. "Schools were not established to teach and encourage the pupil to think," Clarence Darrow wrote of his own nineteenth-century education. "From the first grade to the end of the college course [students] were taught not to think, and the instructor who dared to utter anything in conflict with ordinary beliefs and customs was promptly dismissed, if not destroyed."[44]

This approach to education led to a de facto establishment of Christianity within American public schools. About the time of the Scopes trial, for example, the Georgia Supreme Court dismissed a Jewish taxpayer's complaint against Christian religious exercises in public schools with the observation, "The Jew may complain to the court as a taxpayer just exactly when and only when a Christian may complain to the court as a taxpayer, *i.e.*, when the Legislature authorizes such reading of the Bible or such instruction in the Christian religion in the public schools as give one Christian sect a preference over others."[45] The Tennessee legislature codified a similar practice in 1915 when it mandated the daily reading of ten Bible verses in public schools but prohibited any comment on the readings.[46] This suggestion that constitutional limits on the establishment of religion simply forbad the government from giving preference to any one church denomination reflected a traditional view of religious freedom that dated at least as far back as the great federalist U.S. Supreme Court justice Joseph Story.[47] By the 1920s, however, an increasing number of liberally educated Americans, including leaders of the ACLU, rejected the idea that public education should promote any particular political, economic, or religious viewpoint—even one broadly defined as democratic, capitalistic, or Christian.

The drive to free the American academy from outside political and religious influences began with higher education. Americans originally formed their colleges and universities on the English model, which did not incorporate modern concepts of tenure and academic freedom. At Oxford and Cambridge, for example, faculty members ultimately served under the authority of the Church of England and every college conducted daily Anglican chapel services for students. Similarly, in nineteenth-century America, professors at most public and private institutions of higher education served at the pleasure of the institution's president and trustees, many of whom were ordained ministers, and even Thomas Jefferson's University of Virginia held student chapel services. This did not mean that conservative religious and political ideas held sway on all American campuses—Harvard came under the influence of Unitarianism early in the century, while Oberlin later became famous for its radical egalitarianism and Bryn Mawr for its feminism—but a party line tended to prevail within each institution. Late nineteenth-century populists, progressives, and radicals often accused college administrators of suppressing classroom teaching of alternative economic and political theories. A few highly publicized cases of alleged religious censorship also arose. Coincidentally, the most famous such case took place in Tennessee, where in 1878 the fledgling, southern Methodist-controlled Vanderbilt University terminated the part-time lecturing position of the famed geologist Alexander Winchell for suggesting that humans lived on earth before the biblical Adam. Winchell was an evolutionist, and his firing soon became a cause célèbre in the perceived warfare between science and religion.[48]

The effort to maintain orthodoxy on American campuses encountered increasing resistance around the turn of the century. The historian George M. Marsden linked this development to the rise of pragmatism, flowing from the theories of the French philosopher Auguste Comte. "In Comte's construction of history," Marsden observed, "humans were rising from a religious stage in which questions were decided by authority, through a metaphysical stage in which philosophy ruled, to a positive stage in which empirical investigation would be accepted as the only reliable road to truth."[49] Empirical methods quickly came to dominate academic research in both the sciences and the humanities.

New principles of free academic inquiry and discussion logically followed from these new methods for acquiring knowledge. The Johns Hopkins University and the University of Chicago were founded during the late nineteenth century on the model of German universities, which incorporated basic concepts of professorial tenure and academic freedom. Several existing institutions, including Harvard, Columbia, and Cornell, quickly adopted a similar model. By the 1896 edition of his *A History of the Warfare of Science with Theology in Christendom*, the former Cornell University president Andrew Dickson White could write of the Winchell affair that Vanderbilt had "violated the fundamental principles on which any institution worthy of the name [university] must be based."[50] About this time, the national professional associations for economists, political scientists, and sociologists formed standing committees to investigate individual cases of alleged assaults on academic freedom.

These developments took a decisive turn in 1913, when Lafayette College dismissed the philosophy professor John Mecklin for teaching that social evolution, rather than revealed truth, shaped the development of religious ideas. The American Philosophical Association and American Psychological Association appointed a special committee, chaired by the Hopkins philosophy professor Arthur O. Lovejoy, to investigate the dismissal. Lafayette College defended its action on the grounds that as a denominational institution it could enforce orthodoxy within its curriculum. The committee grudgingly accepted this position, but maintained that "American colleges and universities fall into two classes": either they guaranteed academic freedom or they served as "institutions of denominational or political propaganda," with Lafayette placing itself into the latter class.[51] To give substance to this distinction and thereby promote the rights of faculty members in the former class of institutions, Lovejoy immediately set about forming the American Association of University Professors (AAUP).

With Lovejoy as its first secretary, the AAUP assumed the role of a national guild for university professors. Minutes of the association's organizational meeting reported that members voted "to bring about a merging in a new committee of the committees already created by the economics, political science and sociology associations to deal with the subject of academic freedom."[52] Lovejoy served on this new committee on academic freedom, which presented its General Declaration of Principles at the AAUP's first annual meeting in 1915. Endorsing the distinction emerging from the Lafayette College affair, this document recognized two types of institutions. "The simplest case is that of a proprietary school or college designed for the propagation of specific doctrines prescribed by those who have furnished its endowment," the committee wrote. These institutions, which included many trade schools as well as such church colleges as Lafayette, need not offer academic freedom to their faculty. Institutions receiving support from the government or through appeals to the general public, however, fell into a different category. "Trustees of such institutions or colleges have no moral right to bind the reason or conscience of any professor," the committee asserted, in defiance of traditional practices. To justify this new principle, the committee observed, "In the earlier stages of a nation's intellectual development, the chief concern of educational institutions is to train the growing generation and to defuse the already accepted knowledge." In twentieth-century America, however, "The modern university is becoming more and more the home for scientific research. There are three fields of human inquiry in which the race is only beginning: natural science, social science, and philosophy and religion." In earlier times, the committee added, "the chief menace to academic freedom was ecclesiastical, and the disciplines chiefly affected were philosophy and the natural sciences. In more recent times the danger zone has been shifted to the political and social sciences—though we still have sporadic examples of the former class of cases in some of our smaller institutions."[53] The coming antievolution crusade would refocus attention on this former class.

Despite the prediction that most disputes over academic freedom would involve issues of political and economic ideology, the committee's General Declaration of Principles placed the AAUP on a collision course with the antievolution crusade. Tennessee was again at the center of the storm. Bryan, of course, crusaded against Darwinism in state universities as well as in public schools. After the Kentucky legislature nearly passed an antievolution bill in 1922, the University of Tennessee president Harcourt A. Morgan asked the education professor J. W. Sprowls not to assign Robinson's *Mind in the Making*, which presented an evolutionary view of social progress. Morgan, who included evolutionary concepts in his own biology classes, reportedly told Sprowls that "Tennessee was threatened with legislation such as has recently been proposed in Kentucky, and that it was necessary to 'soft-pedal' the teaching of evolution in the University in order to prevent the enactment of such a law by the Tennessee legislature."[54] Sprowls acquiesced, but soon learned that his annual teaching contract would not be renewed due to deficiencies in his fieldwork, an essential part of his job. Sprowls claimed that he was fired for teaching evolution, however, and soon the campus was in an uproar. Each member of the Tennessee faculty then served under individual one-year contracts and by the time the dust settled, four additional professors were sacked for agitation in defense of Sprowls. At the same time (but for unrelated reasons) the university decided not to renew the contracts of two other instructors, including longtime law professor John R. Neal. AAUP investigators soon descended on Knoxville to investigate the mass firings.

The AAUP investigators criticized the university's handling of the episode. One-year contracts for senior faculty members violated AAUP standards for tenure. The university failed to give timely notice to the four professors fired for defending Sprowls. None of the dismissed teachers received due process. The evidence on charges of religious discrimination was mixed, however. One university official allegedly said, "We are getting rid of a bunch of atheists," but the assertion was demonstrably false and he denied ever saying it. Sprowls continued to cast himself as a martyr to the antievolution crusade, but university officials consistently gave other reasons for their actions that the investigators accepted. "Professor Sprowls' views on evolution," the AAUP report concluded, "were not one of the reasons—certainly not the controlling reason—which led to the decision of the authorities to discontinue his services." The investigators disapproved of Morgan's interference in Sprowls's decision to assign a text on evolution, however, and the continuing public furor in and around Knoxville over the episode helped set the tone for the Scopes trial in nearby Dayton.[55]

Neal's dismissal had an additional impact on the Scopes trial, even though the AAUP investigation found that the action bore no relationship to either the Sprowls affair or teaching evolution. Neal probably missed most of the uproar over Sprowls's dismissal because it occurred in late spring, when Neal typically taught law in Colorado. Indeed, according to his dean, Neal never spent much time on campus—often arriving late (if at all) for class, devoting class time to rambling lectures about current political issues rather than to the course subject matter, and giving all his law students a grade of 95 without reading their exams. The dean also complained about Neal's "slovenly" dress, which later deteriorated into complete disregard for personal appearance and cleanliness. Yet Neal was a loyal Tennessee alumnus who had served two terms in the state legislature and helped secure generous appropriations for the university.[56] After his dismissal, Neal remained in Knoxville trying to establish a rival law school, stirring up a legislative investigation of Morgan, running unsuccessfully for governor, and claiming that he had been sacked for defending the teaching of evolution. When Scopes was indicted in 1925, Neal promptly offered to represent the defendant and ultimately served as local counsel for the defense throughout the case—to the growing frustration of ACLU attorneys in New York.

Although its investigation largely cleared the University of Tennessee of charges that it had suppressed teaching evolution, the AAUP remained concerned about the issue. Its president declared at the time, "Fundamentalism is the most sinister force that has yet attacked freedom of teaching," and the association empaneled a special Committee on Freedom of Teaching in Science to further study this threat.[57] The committee issued its report in December 1924, less than three months before the Tennessee legislature banned teaching evolution. "The last few years have witnessed a revival of the spirit of intolerance which has asserted itself, especially in the opposition to the teaching of evolution," the committee warned. The AAUP would stand against this popular onslaught. "It is, we believe, a principle to be rigidly adhered to that the decision as to what is taught," the committee affirmed, "would be determined not by a popular vote . . . but by the teachers and investigators in their respective fields."[58] During the following summer, several charter members of the AAUP volunteered to go to Dayton to support this position as expert witnesses for the defense at the Scopes trial.

The drive for academic freedom gradually spread from higher education to secondary education—and here the ACLU assumed a leading role. During the 1920s, early ACLU efforts on behalf of pacifism and labor unions in public education blossomed into a broad program to defend academic freedom. Predictably, it began with the ACLU executive committee member Henry Linville, who as head of the New York City teachers union worked closely with the AAUP's first president, Columbia University professor John Dewey. Linville prepared for the ACLU a Tentative Statement of a Plan for Initiating Work on Free-

Speech Cases in Schools and Colleges in the early twenties. The Tentative Statement dealt only with teachers dismissed for expressing their political views outside the classroom, and adopted the AAUP's distinction between publicly supported schools, where the ACLU would intervene, and proprietary schools, where it would not.[59]

ACLU chair Harry F. Ward wanted to reach into the classroom. "The public mind is poisoned at its source when special interests take hold of educational institutions for their own propaganda," Ward shot back in a memorandum. "Most conspicuous are the Lusk laws, recently repealed in New York State, the attempt to rewrite history from a nationalistic viewpoint, and the attacks of the American Legion and other organizations on both the teaching of pacifism and on pacifist students." Although antievolution laws were not yet an express concern, Ward clearly identified "free speech in the class-room" as a potential ACLU priority. "The Union's chief contribution in situations arising in public and private schools," he added, "[is] with protests and with the organization of public opinion." Formal inquiries could be left to professional associations, Ward suggested, but the ACLU should help by "giving the facts national publicity"—a strategy the ACLU would adopt in the Scopes trial.[60]

In mid-1924, the ACLU issued its first public statement on academic freedom. The statement essentially combined Linville's Tentative Statement with Ward's memorandum and identified both men as its co-authors. In it, the ACLU offered to defend the right of public school teachers to free speech both inside and outside the classroom, and explicitly adopted AAUP's conception of academic freedom. Significantly, the new statement added antievolution laws as a "chief issue" of ACLU concern, lumping them together with "Lusk laws" and "history text-book laws" as "cases of propagandists' efforts to distort education." "Whenever any such issue arises in any school or college described in this memorandum, those interested should write or wire the American Civil Liberties Union," the statement concluded. "Aid will be furnished at once either through local correspondents, consulting attorneys or direct from the New York headquarters. In important cases a representative will be sent to the scene of trouble."[61]

To supervise this effort, the ACLU formed an elite Committee on Academic Freedom, which brought longtime ACLU activists Linville, Thomas, Holmes, and Felix Frankfurter together with such prominent educators as Stanford University president emeritus David Starr Jordan. An official release announced that the new committee "will deal with laws restricting teaching, such as those attempting to prohibit the teaching of evolution," and committed the ACLU "to go into each situation promptly, to get the facts before the public, to organize effective protests and to bring to bear national publicity on every local invasion of what we regard as the rights of students and instructors."[62] The release rebroadcast the earlier open offer of free assistance, but the big breakthrough required a narrowly focused appeal from ACLU headquarters in New York.

"I came across a dispatch in a Tennessee newspaper on my crowded desk which was to turn our office topsy-turvy in excitement. It was a three-inch item stating: 'Tennessee Bars the Teaching of Evolution,' " longtime ACLU secretary Lucille Milner later recalled. "I hurriedly clipped the small article and sought Roger [Baldwin]'s advice. 'Here's something that ought to have our attention. . . . What should we do about it?' He glanced over it and saw its import in a flash. 'Take it to the [Executive] Board on Monday,' he said laconically."[63] Baldwin remembered the episode somewhat differently. Milner, whose job included clipping newspapers for reports of civil liberties violations, noticed an article about a proposed Tennessee law. "When we read press reports of what seemed to us a fantastic proposal pending in the Tennessee legislature to make the teaching of evolution a crime, we kept our eye on it," Baldwin wrote. "When the governor signed the bill we at once proffered a press release for Tennessee papers, offering to defend any teacher prosecuted under it. That was the origin of probably the most widely reported trial on a public issue ever to have taken place in the United States."[64]

Baldwin's account rings true. Enactment of the Tennessee antievolution law was a major news story—the first triumph of a four-year national crusade. Only the introduction or consideration of the Tennessee legislation would have been relegated to a small article, especially in a paper from that state. Furthermore, the ACLU closely followed the progress of antievolution legislation in various states throughout the country since the beginning of the crusade and placed them in the context of other restrictions on academic freedom. More than a week before issuing its public offer to assist any teacher in challenging the Tennessee law, the ACLU released a broad survey of restrictions on teaching in schools and colleges. Citing new statutes in seven states to "require daily Bible reading in the schools or forbid employment of radical or pacifist teachers" in addition to the Tennessee antievolution law, the survey concluded that "more restrictive laws had been enacted in the last six months than at any time in the history of the country." In conjunction with releasing this survey, the ACLU announced, "Efforts to get court decisions on all these restrictive laws are being made through Civil Liberties Union attorneys."[65]

The ACLU press release offering to challenge the Tennessee law appeared in its entirety on May 4 in the *Chattanooga Times*, which had opposed enactment of the antievolution statute. "We are looking for a Tennessee teacher who is willing to accept our services in testing this law in the courts," the release stated. "Our lawyers think a friendly test case can be arranged without costing a teacher his or her job. Distinguished counsel have volunteered their services. All we need now is a willing client." Pursuing the story, a *Chattanooga Times* reporter inquired whether city schools taught evolution. "That depends on what is meant by evolution. If you have reference to the Darwinian theory, which, I suppose, was aimed at in the law passed by the Tennessee legislature, it is not," the city school superintendent assured the reporter. "It is recognized by all our teachers that this is a debatable theory and, as such, has no place in our curriculum." Earlier, in making similar assurances regarding his schools, the Knoxville superintendent had noted, "Our teachers have a hard enough time teaching the children how to distinguish between plant and animal life."[66] These urban school officials clearly did not want to test the new law, but midway between these cities enterprising civic boosters in Dayton craved some attention for their struggling community, and accepted the ACLU offer. They got more than they bargained for. Powerful social forces converged on Dayton that summer: populist majoritarianism and traditional evangelical faith versus scientific secularism and modern concepts of individual liberty. America would never be the same again—or perhaps it had changed already from the country that had nurtured Bryan and Darrow in its heartland.

PART II

. . . During . . .

— CHAPTER FOUR —

CHOOSING SIDES

"WHY DAYTON, of all places?" a *St. Louis Post Dispatch* editorial asked in May 1925, "why Dayton?" Local civic boosters adopted this question as the title for a promotional booklet sold during the trial. "Of all places, why not Dayton?" the booklet asked back, "this bowl in the Cumberland holds 'logically, fundamentally and evolutionarily' the amphitheater for a world's comedy or tragedy, whichever viewpoint the spectators may choose."[1] The booklet went on to note that major events happen in obscure places, giving the example of Christ's crucifixion at Calvary, then got down to the serious business of promoting Dayton as a place to live and work, without explaining why "logically, fundamentally and evolutionarily" the trial arose in the town—or why any self-respecting civic leaders would want their community to host such an event. Yet those reasons existed and they helped to explain the entire episode.

Situated midway between Knoxville and Chattanooga in the valley carved by the Tennessee River in the rising foothills of East Tennessee, Dayton lacked both a sense of tradition and confidence in the future. Only a few farmhouses existed in the area at the time of the Civil War, which in 1925 remained a vivid memory for many Tennesseans. The town sprang up in the late 1800s with the coming of the railroads and became the commercial and governmental center for Rhea County. It was part of the so-called New South. Northern money financed laying the rail lines, digging nearby coal and iron mines, and building a blast furnace that attracted hundreds of Scottish immigrants and underemployed Southerners to the new town. Optimistic county officials erected a handsome, three-story courthouse on a spacious downtown square. By linking Dayton to northern markets, the rail lines facilitated the development of commercial farming in the surrounding valley, with Rhea County becoming a major center for strawberry production by the twenties; even though the berry crop flourished and mining continued, the blast furnace went cold. The opening of a hosiery mill early in the new century could not offset the loss of jobs at the furnace. New commercial construction slowed, leaving the downtown with three blocks of one- and two-story storefronts and two sides of the courthouse square undeveloped. Concerned civic leaders actively courted new industry as they watched their town's population dwindle from a peak of about 3,000 during the Gay Nineties to fewer than 1,800 by the time of the Scopes trial.[2]

A New Yorker with some training in chemical engineering, George W. Rappleyea, managed the mines for their northern owners in 1925. Only 31 years old, Rappleyea was described in the *Chattanooga Times* as "a stranger to the south and southern ways."[3] He had drifted away from religion while in college and fully accepted the theory of human evolution. After moving to Dayton, however, he began attending a nearby Methodist church, whose young modernist minister persuaded him that an evolutionist could believe in Christianity. Rappleyea viewed Tennessee's antievolution law with disdain and wrote an indignant letter about it to the *Chattanooga Times* asserting the common modernist line that "John Wesley, the founder of Methodism, . . . advanced the theory of the evolution of man 100 years before Darwin." Upon reading in that newspaper on May 4 about the ACLU offer to help any Tennessee schoolteacher challenge the new law in court, Rappleyea saw a chance to strike the statute—and he set about drawing other townspeople into his scheme.[4]

Rappleyea hurried down to Frank E. Robinson's drugstore with newspaper in hand, or at least that is how the most credible version of this legend goes. Robinson chaired the Rhea County school board, and the soda fountain at his downtown drugstore served as the watering hole for the town's business and professional elite during those days of national Prohibition. "Mr. Robinson, you and [local attorney] John Godsey are always looking for something that will get Dayton a little publicity. I wonder if you have seen the morning paper?" Robinson later recalled Rappleyea asking.[5] Of course Robinson had seen the morning paper but had not noted the ACLU offer. Rappleyea then related his scheme of staging a test case in Dayton and boasted of having connections to the ACLU in New York. Robinson slowly warmed to the idea, as did School Superintendent Walter White, a former Republican state senator who liked the antievolution law but loved publicity for his town even more. Godsey agreed to assist the defense. A few other Daytonians also may have participated at this stage—many later would claim a role—before Rappleyea was confident enough of local support to place his initial call to New York asking whether the ACLU would make good on its offer if Dayton indicted one of its own schoolteachers. Other key participants signed on the next day, when the ACLU accepted the arrangement.

First, Dayton's two young city attorneys, Herbert E. Hicks and Sue K. Hicks, agreed to prosecute the case if a local teacher had taught evolution during the brief period between enactment of the law and the end of the school year. The Hicks brothers (Sue was named for his mother, who died at his birth) were the only persons involved in bringing the case other than Walter White who expressed any sincere concern about teaching evolution, but even they doubted the constitutionality of the antievolution statute. Wallace Haggard, a young Dayton attorney better known for his exploits on the gridiron for Vanderbilt than in the courtroom for clients, volunteered to assist them. The ACLU offered to pay their expenses, but all three declined.[6]

Second, the drugstore conspirators summoned the high school's 24-year-old general science instructor and part-time football coach, John T. Scopes. "Robinson offered me a chair and the boy who worked as a soda jerk brought me a fountain drink," Scopes later wrote. "'John, we've been arguing,' said Rappleyea, 'and I said that nobody could teach biology without teaching evolution.' 'That's right,' I said, not sure what he was leading up to." A chain-smoker, Scopes probably lit a cigarette at this point, if he had not already done so. He then pulled down a copy of Hunter's *Civic Biology* from a sales shelf—the enterprising Robinson also sold public school textbooks—and opened it to the section on human evolution. This was the state-approved text, prescribed for use in all Tennessee high schools. "'You have been teaching 'em this book?' Rappleyea said. 'Yes,' I said. I explained that I had got the book out of storage and had used it for review purposes while filling in for the principal during his illness. He was the regular biology teacher," Scopes recalled. "'Then you've been violating the law,' Robinson said." The school board official then told Scopes about the ACLU offer. Scopes remembered the fateful question: "'John, would you be willing to stand for a test case?' Robinson said. 'Would you be willing to let your name be used?' I realized that the best time to scotch the snake is when it starts to wiggle. The snake already had been wiggling a good long time."[7]

Scopes presented an ideal defendant for the test case. Single, easygoing, and without any fixed intention of staying in Dayton, he had lit-

Dayton trial leaders reenact their original meeting at Robinson's drugstore for press photographers. Seated from left are Herbert Hicks, John Scopes, Walter White, and Gordon McKenzie; standing are constable Burt Wilbur, Wallace Haggard, W. E. Morgan, George Rappleyea, Sue Hicks, and F. E. Robinson. (Courtesy of Bryan College Archives)

tle to lose from a summertime caper—unlike the regular biology teacher, who had a family and administrative responsibilities. Scopes also looked the part of an earnest young teacher, complete with horn-rimmed glasses and a boyish face that made him appear academic but not threatening. Naturally shy, cooperative, and well-liked, he would not alienate parents or taxpayers with soapbox speeches on evolution or give the appearance of a radical or ungrateful public employee. Yet his friends knew that Scopes disapproved of the new law and accepted an evolutionary view of human origins. Not that he understood much about the issue—after all, he taught physics, math, and football, not biology—but he was a student at the University of Kentucky when that institution's president led the fight against antievolution legislation in the Bluegrass State, and he admired the president's courage. Furthermore, Scopes's father, an immigrant railroad mechanic and labor organizer, was an avowed Socialist and agnostic who, as the *Chattanooga Times* reported, "could talk long and loud against the political and religious system of America."[8] John Scopes inclined toward his father's views about government and religion, but in an easygoing way. Indeed, he liked to talk about sports more than politics and occasionally attended Dayton's northern Methodist church as a way to make friends. This was their defendant, an establishment's rebel who would test the law without causing trouble. "Had we sought to find a defendant to present the issue," ACLU counsel Arthur Garfield Hays later confided, "we could not have improved on the individual."[9]

Although Rappleyea and Robinson pressed the young teacher to accept the challenge, Scopes could have refused. Sue Hicks stood at his side; the two young men were close friends. "After we had discussed that possibility for a while, Scopes said he would be glad to do it, and I said I wouldn't mind to prosecute him," Hicks reported. Rappleyea then called over a nearby justice of the peace, swore out a warrant for Scopes "arrest," and handed it to a waiting constable to "serve" on the accused.[10] After Scopes left for a game of tennis, Rappleyea wired the ACLU in New York while Robinson called the *Chattanooga Times* and *Nashville Banner*. Walter White, for his part, hailed the local stringer for the *Chattanooga News* with the words, "Something has happened that's going to put Dayton on the map!"[11] The show had begun.

The next day, a front-page article in the *Banner* carried the story. "J. T. Scopes, head of the science department of the Rhea County high school, was . . . charged with violating the recently enacted law prohibiting the teaching of evolution in the public schools of Tennessee. Prof. Scopes is being prosecuted by George W. Rappleyea, manager of the Cumberland Coal and Iron Co., who is represented in the prosecution by S. K. Hicks," the newspaper reported. "The defendant will attack the new law on constitutional grounds. The case is brought as a test of the new law. The prosecution is acting under the auspices of the American Civil Liberties Association [*sic*] of New York, which, it is said, has offered to defray the expenses of such litigation."[12] The Associated Press picked up this article and transmitted it to every major newspaper in the country.

Anyone capable of reading between the lines of this article could see that *Tennessee v. Scopes* was not a normal criminal case. Enmity typically characterized the relationship between the ACLU and public officials during the twenties. Prosecutors simply did not act under ACLU "auspices." The ACLU would never "defray" prosecution expenses. School officials rarely publicized criminal charges filed against one of their own teachers. Everything about this case appeared upside-down, and no one seemed to care about upholding the law. Of course, the statute itself was unusual. From the outset, Bryan counseled against including a penalty provision, and Governor Peay predicted that it would never be enforced. As if taking a cue from these leaders, other Tennessee communities passed over the opportunity for what the *Nashville Tennessean* denounced as "cheap publicity."[13] In Dayton, however, civic leaders all but manufactured a test case, then bragged about doing so. Even the ACLU failed at first to appreciate the uniqueness of the situation; its initial press release predicted "a conviction in the trial" and stressed its "arrangements" for the appeal.[14] Clearly, the ACLU anticipated a typical constitutional test case in which most of the serious lawyering occurred at the appellate level. Daytonians, of course, had other ideas.

These developments did not necessarily reflect on Tennessee as a whole. The offer to stage the trial came from New York, as did Rappleyea. Scopes hailed from Salem, Illinois—Bryan's hometown. Even Sue Hicks was new to Dayton, and all three men soon moved on. The town itself was new—and fundamentally disconnected from its state and region. East Tennessee was the only major Republican enclave in the entire South. Bryan swept every southern state during his three runs for the presidency, but never carried Rhea County. Peay typically trailed there as well. Local politicians did not owe allegiance to the Commoner or the governor. Religious differences existed too: Tennessee and the South were essentially Baptist, whereas Dayton was mostly Methodist—and, as H. L. Mencken joked during the trial, "a Methodist down here belongs to the extreme wing of liberals." Furthermore, a relatively high percentage of Dayton residents did not belong to any denomination; indeed, the town's Masonic lodge claimed more adult male members than any local church. But Dayton was not a hotbed of modernism. An informal survey conducted during the trial found that 85 percent of persons attending Dayton churches professed to believe the Bible literally. More than likely, however, few could have listed the basic tenets of fundamentalism; it was not that sort of town.[15]

Journalists covering the trial commented on Dayton's distinctiveness from the outset. None found rancor. Even the hypercynical Mencken wrote in his first report from Dayton, "The town, I confess, greatly surprised me. I expected to find a squalid Southern village, with darkies snoozing on the houseblocks, pigs rooting under the houses and the inhabitants full of hookworm and malaria. What I found was a country town full of charm and even beauty." Dayton was too new to be run-down and attracted fewer African Americans than many similarly sized manufacturing towns of the North and Midwest.[16] "Nor is there any evidence in the town of that poisonous spirit which usually shows itself where Christian men gather to defend the great doctrines of their faith," Mencken added. "The basic issues of the case, indeed, seems to be very little discussed at Dayton. What interests everyone is its strategy."[17] Within a few days, Mencken realized that their primary interest was in good publicity and quietly slipped out of town before his increasingly caustic reports got him in too much trouble.

Other Tennesseans did not appreciate Dayton's publicity "stunt," as the *Chattanooga Times* termed the Scopes trial. Vanderbilt University's renowned humanist Edwin Mims described the antievolution law as "disturbing" and the pending trial as "deplorable." Governor Peay pointedly refused to attend the proceedings, despite a request from town officials. "It is not a fight for evolution or against evolution, but a fight against obscurity," Congressman Foster V. Brown of Chattanooga complained. Knoxville congressman J. Will Taylor added, "The Dayton trial will be a travesty."[18]

Every major newspaper of the state—even those opposed to the antievolution law—criticized Dayton for staging the trial. "Apparently the 'booster' element in Dayton have with questionable wisdom and taste, seized on this as an opportunity to get widespread publicity for their city," a *Nashville Tennessean* editorial observed, "evidently proceeding on the doubtful theory that it is good advertising to have people talking about you, regardless of what they are saying." The *Knoxville Journal* commented, "The major actors on the Dayton stage [are] there for publicity and don't care three straws what may be decided in the court." Still smarting from passage of the antievolution law, the editor of the *Chattanooga Times* denounced the Scopes trial as "a humiliating proceeding" and claimed that "every lawyer in the state is holding his head in shame." The most charitable comment came from the *Nashville Banner*, which simply observed, "Dayton could not have overlooked such an opportunity to secure front page advertising space throughout the civilized world." Fittingly, the *Banner* later proposed that the entire state horn in on this opportunity by having business leaders charter a train to give visiting journalists a tour of "progressive" Tennessee. "It should be made as fine a train as the state ever saw," the editor envisioned. Of course, it never materialized.[19]

Tennesseans living outside Rhea County tended to foresee nothing but bad publicity coming from what the *Chattanooga Times* called "the Dayton serio-comedy."[20] This was the main reason why the *Banner* wanted to show other parts of the state to visiting journalists. Those proud of their state's antievolution statute feared that the upcoming trial would discredit it and Tennessee; those embarrassed by it feared that the upcoming trial would heap further ridicule on their state.

Still suffering from the ravages of the Civil War and the humiliation of Reconstruction, Southerners were conscious of their national image and sensitive to any perceived slight. Fourteen southern states, including Tennessee, had just opened a promotional exposition in New York City. "The South has suffered from reports of illiteracy among its population, of poverty among its people, of backwardness in business," the mayor of New York observed at the time. "This exposition is a progressive step . . . to tell the rest of the country some of the facts about Southern progress."[21] Tennessee editorialists chronicled the response to the exposition. "The New York newspapers as a rule are commenting most kindly," a *Chattanooga Times* editorial noted, but it took extreme exception to a *Herald-Tribune* comment that, "while the south has advanced industrially, it has not recovered intellectually."[22] After quoting from several laudatory reports in the New York press, a *Nashville Banner* editorial added, "The world is taking note of the South." In the same issue, however, the *Banner* carried the initial report from Dayton about Scopes's "arrest."[23] The world *would* take note—but of the later event, and it was not the type of notice that the *Banner* wanted for the state; it simply reinforced stereotypes about intellectual backwardness and eclipsed any "facts about Southern progress" coming from the exposition.

Daytonians pushed on without regard to their critics. At a preliminary hearing on May 9, the county's three justices of the peace formally held Scopes for action by the August grand jury, in the meantime releasing him without bond. Knoxville's eccentric law professor, John Randolph Neal, who ran his own proprietary law school following his dismissal from the University of Tennessee faculty and failure to unseat Peay in the 1924 gubernatorial primary, appeared for the defense at this hearing, along with Godsey. Neal had driven to Dayton a few days earlier without invitation and presented himself to Scopes with the words, "Boy, I'm interested in your case and, whether you want me or not, I'm going to be here." The ACLU had not yet made arrangements for legal representation and in any event Scopes needed local counsel, so the relationship stuck.[24] At the hearing, Neal readily conceded that Scopes taught about human evolution but denied that it conflicted with the biblical account. "Legislative enactment can not make it so," he asserted. This would remain Scopes's principal argument throughout: the majority, acting through the legislature, cannot define the tenets of science or religion for individual public school teachers or students. "We regard it as equally un-American, and therefore unconstitutional, whether it is kingly or ecclesiastical authority or legislative power that would attempt to limit the human mind in its enquiry after truth," Neal explained.[25]

Trial promoters welcomed Neal as the first noted outsider to join the proceedings but cried foul when he suggested transferring the case to Knoxville or Chattanooga, which could provide more dignified facilities for the event and adequate accommodations for visitors. The *Chattanooga News* even tried to instigate a new case in its city in the event that the transfer fell through. Daytonians responded by threatening to boycott Chattanooga merchants and preparing their town for the trial.[26] The leading civic association, the Progressive Dayton Club, formed a Scopes Trial Entertainment Committee to arrange suitable trial facilities and visitor accommodations. "A strong following has been mustered for erecting a gigantic tent," the *Nashville Tennessean* reported. "Others favor placing a roof over the baseball park and there are those who stand solidly behind a plan to fill every inch of the courtroom with seats, place benches on the huge lawn and use loud speakers." Proponents of the third option stressed that Dayton already had an unusually large courtroom—the second largest in the state—and they ultimately prevailed. Housing the thousands of expected visitors posed a trickier problem because Dayton had only three hotels, with a combined total of two hundred rooms. The committee provided for further accommodations though a card index of rooms in private homes and requisitioning army tents and cots through East Tennessee's powerful representative in Congress, future Secretary of State Cordell Hull. To clinch the case for Dayton, the district judge, acting with the consent of both prosecution and defense, called a special session of the grand jury for May 25 to indict Scopes before any other town could steal the show.[27]

Carried away with these developments, trial promoters invited the British evolutionist and writer H. G. Wells to present the case for evolution. "I am sure that in the interest of science Mr. Wells will consent," Rappleyea told reporters. Of course, Wells, a popular writer and speaker—not a lawyer—summarily dismissed the idea, although he did take up the cause against Bryan and antievolutionism in articles and addresses. The invitation, however, suggested that Daytonians envisioned the upcoming trial more as a public debate around Scopes than as a criminal prosecution against him. Indeed, it had all the trappings of a summer Chautauqua lecture series, then a popular form of education and entertainment in communities throughout America. In mid May, their vision began crystallizing into reality when William Jennings Bryan—a top draw on the Chautauqua circuit—volunteered his services for the prosecution.[28]

Strictly speaking, it made no more sense for Bryan to appear as an attorney for this case than for Wells to do so. The Commoner had not practiced law for more than thirty years. Traditionally, a Washington or a Lincoln served as the model for American political leaders—a planter or an attorney elected by the people to political office, who then returned to private life after public service. Bryan followed a newer model—one that would become common later in the twentieth century. He adopted a series of political causes, from monetary reform in the 1890s to antievolution legislation in the 1920s, and championed them full time win or lose. Through lecture fees and book contracts, Bryan earned far more money speaking and writing about these causes out of office than he ever earned from his government salary as a member of Congress or the cabinet. He so loved the spotlight and passionately believed in his causes that returning to the practice of law held little attraction for him.

Of course, success as a lecturer and author required a steady stream of popular causes appealing to broad audiences, and Bryan generated them through his distinctive combination of left-wing politics and right-wing religion. Although the mainstream press often scoffed at this antiestablishment mix, he continued to make headlines after such fellow progressives as Theodore Roosevelt, Woodrow Wilson, and Robert La Follette passed from the scene. Commenting on this, an editorial cartoon during the Scopes trial showed a defiant Bryan sitting on the front page for thirty years, with the notation: "You can't laugh that off!" Others had noted the Commoner's ability to retain popular influence despite ridicule, but Bryan called this cartoon "the best of its line" and asked the cartoonist for the original.[29] Fully cognizant of his role, Bryan did not enter the Rhea County Courthouse as a lawyer prosecuting a case before a small-town jury but as an orator promoting a cause to the entire nation. Daytonians wanted it that way.

Sensing a prime opportunity to gain publicity for his cause, Bryan jumped at the chance to join the prosecution. By coincidence, the World's Christian Fundamentals Association was meeting in Tennessee at the time of Scopes's initial arrest, with Bryan as its featured speaker. As a means to maximize its influence, the WCFA regularly met in conjunction with major church conferences and that year picked the Southern Baptist's annual assembly in Memphis. Even though the time and place of its meeting had nothing to do with the state's new antievolution law, the WCFA wasted no time in adopting a resolution commending Tennessee for "prohibiting the teaching of the unscientific, anti-Christian, atheistic, anarchistic, pagan, rationalist evolutionary theory."[30] The Tennessee connection assured that antievolution lawmaking and the pending lawsuit would be major topics of conversation at the meeting—and helped to attract Bryan, a Presbyterian lay leader, to the Baptist-dominated event. He stressed both topics in his address.

Bryan's address repeated the three main points of his standard argument for antievolution laws: evolution theory lacked scientific proof; teaching it to school students undermined their religious faith and social values; and most important, that the "Bible-believing" majority should control the content of public school instruction. To this he added two new warnings. First, widely publicized ridicule of the Tennessee law was eroding public support of such statutes elsewhere. "Peo-

Editorial cartoon commenting on Bryan's ability to make headlines and create issues. (Reprinted with permission from the *Columbus Dispatch*)

ple who hold the Bible dear should make themselves heard. Recently a lot of [University of Tennessee] students ridiculed the Legislature of your state for passing a bill to prohibit teaching evolution," Bryan observed. "I saw large [newspaper] space given the ridicule but small space given to the noble act of Governor Peay in signing the bill." Second, court challenges posed a further threat. "I notice that a case is on the docket for trial involving the evolution statute of your state. I certainly hope it will be upheld. It ought to be," he concluded.[31] Bryan quickly perceived the pending trial as a vehicle for making himself heard—a "battle royal" in defense of the faith, as he would call it.

After Bryan left the WCFA meeting, leaders of the association who stayed on in Memphis for the Baptist conclave grew increasingly concerned about the upcoming trial. It became readily apparent—at least to those reading Tennessee newspapers—that the ACLU and local civic leaders were staging the event and that no one in Dayton cared much about upholding the law. Indeed, to counter this impression, Sue Hicks issued a press release affirming his commitment to defend the law's validity, and Rappleyea turned the formal role of prosecutor over to Walter White, who held more conservative religious views. Still, the cards appeared stacked against the law. Furthermore, in response to pleas for tolerance, delegates at the Baptist conclave in Memphis overwhelmingly defeated a motion by fundamentalists to add an antievolution plank to the denomination's statement on faith. During his address to the WCFA meeting in Memphis, Bryan described "the south as the bulwark of Fundamentalism, where it would take its last stand if brought to bay." Now even the Southern Baptist Convention rejected a call for laws against teaching evolution. Thrown on the defensive, William Bell Riley and other WCFA leaders sent Bryan a telegram on May 13 asking him to appear on the association's behalf at the Scopes trial. Local attorneys could not be trusted to defend the statute, and antievolutionists desperately needed some sort of victory.[32]

Getting wind of the WCFA request a day later, the *Memphis Press* wired the Rhea County prosecutors, "Will you be willing for William Jennings Bryan to aid the state in prosecution of J. T. Scopes."[33] Sensing an opportunity to secure a top star for Dayton's show, Sue Hicks wired back an affirmative reply and dashed off a letter to Bryan: "We will consider it a great honor to have you with us in the prosecution."[34] Bryan already had publicly accepted the WCFA offer by the time

Hicks's letter reached him nearly a week later at the Presbyterian General Assembly's annual meeting in Columbus, Ohio, where modernists and moderates within the denomination joined forces to rout the Commoner's fundamentalist faction. Not only did the assembly reject the fundamentalist candidate for moderator and a resolution against teaching evolution, but Bryan lost his post as vice-moderator. Eager to regain the offensive, Bryan scribbled a note to Hicks on hotel stationery, "I appreciate your invitation [and] shall be pleased to be associated with your forces in the case." In the margin, he added, "I shall, of course, serve without compensation."[35] In a stroke, the ACLU lost control of what it initially conceived as a narrow constitutional test of the statute. With Bryan on hand, evolution would be on trial at Dayton, and pleas for individual liberty would run headlong into calls for majority rule.

The ACLU's plan for a narrow test case promptly suffered a second setback when Clarence Darrow stepped forward to duel Bryan. Darrow first learned of the pending trial while in Richmond, Virginia, to address the annual meeting of the American Psychological Association on his ideas about the lack of individual criminal responsibility. The address followed Darrow's sensational triumphs in the Leopold–Loeb trial and in the case of a suburban Chicago riding master who murdered his wife. The defendants confessed to the acts in both cases, but Darrow saved them from the death penalty by invoking psychological determinism—quoting from Omar Khayyam in the later case, "We are but the puppets in the games we play."[36] These two trials became the talk of the nation and, at age 68, restored Darrow to prominence as America's leading defense lawyer. H. L. Mencken covered the Richmond address, and the two discussed whether Darrow should defend Scopes; but the aging attorney had just announced his retirement and let the matter pass. The ACLU would not want his help anyway, Darrow surmised, because his zealous agnosticism might transform the trial from a narrow appeal for academic freedom to a broad assault on religion. Furthermore, ever since the Leopold–Loeb trial, Bryan had used Darrow's arguments about the psychological impact of the defendants' study of Nietzsche as a prime example of the need to stop teaching evolution. As the ACLU later assured its many liberal religious supporters, it did not want Darrow anywhere near Dayton.

When Bryan jumped in, however, Darrow could no longer restrain himself. "At once I wanted to go," he later acknowledged. "To me it was perfectly clear that the proceedings bore little semblance to a court case, but I realized that there was no limit to the mischief that might be accomplished unless the country was aroused to the evil at hand."[37] Darrow at that time was in New York consulting with Dudley Field Malone, a swank international divorce lawyer with a passion for radical causes. Malone once served as Bryan's assistant at the State Department and still harbored resentment against his former boss from those days. Darrow and Malone wired Neal and simultaneously released the contents of their telegram to the press. "We have read the report that Mr. William Jennings Bryan has volunteered to aid the prosecution," the telegram noted in language clearly intended for public broadcast. "In view of the fact that scientists are so much interested in the pursuit of knowledge that they can not make the money that lecturers and Florida real estate agents command, in case you should need us, we are willing, without fees or expenses, to help the defense of Professor Scopes."[38]

Adding a sharp edge to the basic argument for individual freedom, Darrow and Malone thus characterized the case as innocent, truth-seeking scientists versus an oppressive, fundamentalist huckster. Darrow was not about to let Bryan set the tone for this debate and the press knew it. Noting that the Commoner's presence "brings the trial of J. T. Scopes into the limelight of a national event," Joseph Pulitzer's *St. Louis Post-Dispatch* commented at the time, "Now for a fitting foeman for Mr. Bryan to speak for evolution—Clarence Darrow, for instance—and we may have a debate that would drag the country out of its doldrums of steadily improving business prospects and corresponding mental lethargy."[39]

Caught off guard by the public offer from Darrow and Malone to Neal, the ACLU never regained control of events. The impulsive and independent-minded John Neal further complicated matters by publicly accepting the offer on Scopes's behalf without consulting the ACLU. To counter Bryan, the ACLU had two former presidential nominees in mind as alternative choices to lead the defense, the Democrat John W. Davis and the Republican Charles Evans Hughes, but neither would serve on a team that Darrow would inevitably dominate. "Even after the selection was made by Scopes, we did the best we could to undo it," ACLU Associate Director Forrest Bailey later explained in a confidential letter to *New York World* editor Walter Lippmann. "We actually had Darrow and Malone right here in our office in an effort to persuade them they did not belong. But here is where another element entered into the scheme of things which we could not control." Without informing the ACLU, Rappleyea asked former Secretary of State Bainbridge Colby to join the defense team. The sometimes erratic Colby, who had jumped back and forth between political parties and was considering a run for the 1928 Democratic presidential nomination, initially agreed to serve with Darrow and Malone. Now, to get Colby, who ACLU leaders found acceptable, they had to retain the others. "Time was pressing and no other respectable eminent counsel was agreeable," Bailey concluded in his letter to Lippmann.[40]

The ACLU made one more attempt to displace Darrow. It occurred in early June, when Scopes, Neal, and Rappleyea went to New York to confer with the ACLU and meet the press. Felix Frankfurter came down from Harvard Law School for the strategy meetings and, with Bailey and ACLU executive director Roger Baldwin, tried to talk Scopes into choosing other counsel. Three New York lawyers with close ties to the ACLU, Arthur Garfield Hays, Samuel Rosensohm, and Walter Nells, also participated in these closed-door meetings—with only Hays backing Darrow.[41] "The arguments against Darrow were various," Scopes later wrote, "that he was too radical, that he was a headline hunter, that the trial would become a circus." But Malone also had a chance to lobby Scopes in New York, and the defendant stuck by Darrow. Facing a criminal prosecution, Scopes wanted an experienced defense lawyer rather than a dignified constitutional attorney. "It was going to be a down-in-the-mud fight," he recalled, "and I felt that situation demanded an Indian fighter rather than someone who graduated from the proper military academy."[42]

Darrow had the right experience and reputation for the job. "[We] adjusted ourselves as gracefully as we could to the presence of Darrow among counsel," Bailey later wrote on behalf of ACLU leaders, but they never liked the idea.[43] Baldwin pointedly refused to participate further and thereby missed his organization's most famous trial. More than a year later, ACLU Counsel Wollcott H. Pitkin confided to Frankfurter, "In my belief, a great mistake had been made at the start in accepting the services of Mr. Darrow, thereby allowing fundamentalists to present the issue as a clash between religion (represented by themselves) and anti-religion" represented by Darrow.[44] Everyone smiled for the press at the end of the New York meetings, however, when Neal announced a defense team consisting of Darrow, Colby, Rosensohm, Malone, and himself. Colby later dropped out, and Hays replaced Rosensohm as the sole ACLU representative in Dayton. Darrow had stolen the leading role.

Despite naming Malone to the defense team, ACLU leaders hoped that he would remain in New York during the trial. Neal's announcement suggested as much by stating, "Mr. Malone has generously offered to take any assignment in the case." Apparently relying on comments by Bailey and Frankfurter—both publicly took this position—the *New York Times* reported, "It is said that Mr. Rosensohm and Mr. Malone would probably have charge of looking up references." It was dangerous enough sending a professed agnostic to Dayton, Frankfurter reasoned, but it was too much to add a divorced Irish Catholic. At least Darrow had a folksy manner. Malone was a pompous city slicker. "I will not be the goat," Malone shot back in a widely publicized statement. "I am accustomed to letting my clerks look up references for me." Again the ACLU backed down, in what the *Chattanooga Times* described as "another victory for those who want to introduce a dramatic setting into the case, the 'jazz' factor, as it were."[45]

Darrow set the tone for the case almost immediately. One day after Neal accepted his offer of help, the Chicago attorney redoubled his efforts to put Bryan on the defensive. "Nero tried to kill Christianity with persecution and law. Bryan would block enlightenment with law," he declared to the press. "Had Mr. Bryan's ideas of what a man may do towards free thinking existed throughout history, we would still be hanging and burning witches and punishing persons who thought the earth was round."[46] If Darrow had his way, Bryan would replace Scopes in the role of the accused. It was a simple theme and one Darrow kept reiterating until he hounded his target into the witness chair at Dayton. The Great Commoner—the self-proclaimed voice of majority rule and religiously motivated progressive reform—would personify the threat to individual liberty in America. Darrow characteristically presented this threat as emanating from religious bigotry, making antievolution laws appear particularly ominous, whereas the ACLU previously had encountered such a threat principally from superpatriotism during the war and cutthroat capitalism thereafter, Bryan having stood for freedom in both instances. Thus, many ACLU supporters questioned the substance of Darrow's attack on Bryan and religion, as well as its strategic effect.[47]

After thirty-five years in partisan politics, Bryan could defend himself in a public debate. He brushed aside Darrow's initial personal attacks with sharp remarks of his own. "Darrow is an atheist, I'm an upholder of Christianity. That's the difference between us," Bryan observed during his next press conference. "I never attempt to answer atheists, or those who argue for the sake of arguing, so will make no reply to Mr. Darrow's attack." He did, however, seek to refocus the debate squarely onto his terra firma of majority rule. "The real issue is not *what* can be taught in public schools, but *who* shall control the education system," the Commoner asserted. "If the people are not to control the schools, who shall control them? Only two other kinds of control have been suggested:" by scientists or by individual teachers. He dismissed the former as undemocratic and the latter as unrealistic. "The absurdity of this [latter] suggestion becomes apparent when the liberty is employed to teach anything that the taxpayers really object to," Bryan explained, such as anti-American or antireligious slander. Teaching evolution apparently fit into the second category. Darrow later claimed that Bryan shifted his focus to the issue of majority rule only after it became apparent that the defense would win any courtroom contest over evolution, but at the outset the Commoner predicted, "The case may be determined without any discussion whatsoever of the merit of evolution."[48]

Although Bryan took advantage of the widespread interest aroused by the upcoming trial to lecture and write about the scientific and moral failings of Darwinism, he never said that he would raise these issues at the trial itself. Having established them through the legislature, he had little to gain by litigating them in court simply to uphold the legislation. "The disgrace is not the Tennessee law," Bryan declared in a typical pretrial speech, "the disgrace is that teachers . . . should betray the trust imposed on them by the taxpayers" by violating the law. Arguing for popular control over public education gave Bryan the legal and logical upper hand in the Scopes case, and he held firmly to that position until he had all but grasped victory. Even the otherwise hostile *New York Times* agreed with him on this narrow point. When Bryan promised "a battle royal between the Christian people of Tennessee and the so-called scientists," it was over which of them should control Tennessee public education, not the truth of evolution per se. "It would be ridiculous to entrust the education of children to an oligarchy of scientists," he maintained.[49] At the time, Darrow would have great difficulty challenging this position in either a court of law or the court of popular opinion.

With an all-star cast assembling, Daytonians pressed forward with their preparations for the trial. Newspapers estimated that up to 30,000 visitors would descend on Dayton for the confrontation between Bryan and Darrow. Although the press gave no basis for this figure, which overestimated the actual crowd by a factor of ten, townspeople planned accordingly. Town officials asked the Southern Railway to schedule extra passenger trains to and from Chattanooga on the days of trial. They requested that the Pullman Company park sleeping and dining cars on nearby rail sidings to accommodate the added numbers. They even petitioned the governor to call up the state militia to control the expected crowds, but had to settle for hiring six extra policemen from Chattanooga for this task.

Townspeople embraced the unfolding affair. "Previous to the consciousness that Dayton was gaining notoriety through Scopes, Rappleyea et al. there was a lot of bickering and dispute"—including assaults on some trial proponents, one journalist reported from Dayton, three days after Bryan and Darrow volunteered to participate. "But now that the trial has been put into the advertising class, monkey has become the most popular word in Dayton's vocabulary." Main Street merchants decorated their shops with pictures of apes and monkeys. One billboard featured a long-tailed primate holding a bottle of patent medicine; another pictured a chimpanzee drinking a soda. The constable's motorcycle carried a sign reading "Monkeyville Police," while a delivery van bore the words "Monkeyville Express." Merchants toned down their displays after the Progressive Dayton Club passed a resolution "condemning the frivolous attitude being taken toward the evolution case by certain elements of the Dayton population," but the fountain at Robinson's drugstore still offered "simian" sodas and stray monkeys continued to appear in shops around town. At the same time, however, the club voted to raise a $5,000 advertising fund to promote business development during the trial. "Since Dayton had found her way into the headlines all over the country," one club member commented, "I can not see why Dayton should not reap the benefits."[50]

Darrow's entry into the case aroused some local protests, however.

Countless Americans never forgave Darrow for his role in the Leopold–Loeb trial. "The fact that [this] and others of his cases were personal victories for himself does not by any means connote that they were also victories for the majesty and efficacy of the law," the Memphis *Commercial Appeal* had commented earlier.[51] Others distrusted Darrow due to his militant agnosticism; Malone was less well known, but rumored to be a Socialist. A group of prominent Daytonians asked Neal to decline aid from Darrow and Malone. Scopes's original attorney, John Godsey, agreed with this position and soon bowed out of the case. Even Rappleyea, who hoped that the trial would promote a modernist Christian view of evolution rather than a materialistic one, did not want Darrow. "Dr. Neal accepted Darrow's help in the case," he told the press, "I wish he had not." After asking numerous Daytonians about the matter, however, a *Chattanooga Times* reporter concluded that "the big majority look at this feature of the case as purely professional, and are ready to congratulate Judge Neal that he has succeeded in adding enormously to the advertising value of the trial by securing these two men of international reputation." Scopes clearly agreed, telling the press at the height of the controversy, "I would certainly be an imbecile not to accept them."[52]

Neal stuck by Darrow for the time being, but differences in their approach to the trial surfaced almost immediately, and each later conspired to remove the other from the case. Whereas Darrow approached it as the culmination of his lifelong struggle against religious intolerance, Neal viewed it as a chance to relitigate his dismissal from the University of Tennessee faculty. "The question is not whether evolution is true or untrue," Neal observed at the outset, "but involves the freedom of teaching, or more important, the freedom of learning."[53] This widely reported comment seemed to unleash a flood of letters and comments to Neal about the theory of evolution and its relationship to religion. "Even the Negro waiters in restaurants and the hotel bell-hops want to give me their views on evolution," he soon complained to the press in a comment that betrayed a consciousness of status consistent with his references to Scopes as "boy."[54] Nevertheless, Neal maintained his focus on academic freedom. Losing all sense of perspective on this topic, he asserted just before trial, "Scopes' case involves the most vital issue—human freedom—that has ever risen in America, transcending the fundamentals underlying the Civil War."[55]

Neal discussed his legal strategy with the press on the eve of the special grand jury proceeding against Scopes. "The fight will continue along the lines I have outlined," he stated, "namely, the lack of power upon the part of the legislature to limit the inquiry of the truth in our high schools and universities." Neal made a bow to those pressing him to defend the theory of evolution by adding, "While, as I have stated most emphatically, this is not a question of the truth or falsity of the Darwin theory, we think it advisable that the judge and jury, in order to secure a proper understanding of the law, should be enlightened in regard to the doctrine of evolution." Godsey elaborated on this point by noting, "Our idea of evolution is that it is absolutely compatible and consistent with the story of creation in the Bible."[56] If so, the defense reasoned, teaching evolution would not violate the statute. Of course, this did not correspond with Darrow's materialistic view of evolution—and the coherence of the defense case suffered accordingly. Neal and the ACLU would fight primarily for academic freedom and secondarily for a broader understanding of evolution and religion, themes that overlapped but never fully coincided with Darrow's agenda. For his part, Scopes declined to give further interviews to the press, thereafter making only brief public appearances in controlled settings.

The grand jury heard none of this when it met on the morning of May 25, only the prosecution's prima facie case against Scopes. Rhea County was in Tennessee's eighteenth judicial district; therefore the task of presenting this case fell on Tom Stewart of Winchester, Tennessee, the district's no-nonsense attorney general. Stewart would become the only lawyer on either side to gain the universal respect of people associated with the case and later served Tennessee in the U.S. Senate. The Hicks brothers and Haggard assisted him throughout, as did Gordon McKenzie, a young Dayton attorney who supported the antievolution law on religious grounds. "I insist that the teaching of evolution in public schools is detrimental to public morals, storms the very citadel of our Christian religion, repudiates God and should not be permitted," McKenzie told the press shortly before the grand jury met.[57] Stewart took a different approach before the grand jury, however. He simply introduced the assigned textbook into evidence through the testimony of the school superintendent, read passages about human evolution from the text, and called on several students to testify that Scopes used the book to teach about human evolution.

Even though Scopes taught both boys and girls, only his male students appeared in court. In all, seven students attended the grand jury session, but Stewart called on only three of them to testify. His entire case took less than an hour.

Scopes had urged the students to testify against him, and coached them in their answers. They did not appear to understand the concept of human evolution, however, when questioned by reporters before the hearing. None of them knew the meaning of "anthropoid ape," though several recalled Scopes talking about Tarzan of the Apes. "I believe in part of evolution, but I don't believe about the monkey business," one boy blurted out. "This was the crime established in the jury room," one reporter lamented.[58]

The presiding judge, John T. Raulston from near Chattanooga, pushed for an indictment. He clearly wanted the trial to proceed and looked forward to his role in it. Earlier, he had expedited the case by reconvening the grand jury in special session to indict Scopes, even though state bar officials questioned the legality of such a procedure. Then, before the grand jury reconvened, he offered to move up the trial date as well. Raulston's judicial circuit sprawled across more than a half dozen rural counties in southeastern Tennessee, and he was not scheduled to hold court in Rhea County until autumn. Yet he publicly announced his willingness to begin the Scopes trial as early as mid June. Finally, Raulston all but instructed the grand jury to indict Scopes, despite the meager evidence against him and the widely reported stories questioning whether the willing defendant had ever taught evolution in the classroom.

Following the prosecution's presentation to the grand jury, Raulston read aloud the antievolution statute and the entire first chapter of Genesis. He then formally charged the jury: "If you find that the statute has been thus violated you should indict the guilty party promptly. You will bear in mind that in this investigation you are not interested to inquire into the policy or wisdom of this legislation." After conceding that the crime carried a low penalty, Raulston volunteered his opinion, "I would regard a violation of this statute as a high misdemeanor. And in so declaring I make no reference to the policy or constitutionality of the statute, but to the evil example of the teacher disregarding constituted authority in the presence of those whose thought and morals he is to direct." So admonished, the all-male jury returned an indictment before noon. "There was not a woman in the courtroom," one journalist noted. "Evidently the day of women has not arrived here." The jury's foreman, a retired mine manager, later told reporters that he personally accepted the theory of evolution but felt compelled to indict Scopes for violating the law.[59]

Raulston's eagerness to push the case reflected two factors. As a politician and elected office holder, Raulston craved publicity. As a conservative Christian and lay Methodist minister, he felt a deep sense of purpose in his work. Both factors were evident at the grand jury session when, after arriving late, he took time to pose for photographers and to give a statement to the press. "It has always been and is now one of the great passions of my life to ascertain the truth about all matters, especially relative to God," he told reporters at this time, "but I am not so much exercised over the question as to whence man cometh as I am to whither he goes."[60] Accordingly, he ended up preaching to Darrow at the trial about the Christian plan for salvation but carefully refrained from expressing any views on human evolution. He apparently felt called by God to preside over this trial and would not let the opportunity slip through his hands.

Raulston concluded the special court session by scheduling the trial to begin on July 10. "I have set a date when all universities and schools will be through their terms of school in order that scientists, theologians and other school men will be able to act as expert witnesses," he explained. Both parties agreed to the early start. Raulston's grandiose expectations for the trial lost touch with reality, but help to explain why he granted such latitude to the litigants. "My suggestion is that a roof be built over a large vacant lot," he explained to assembled reporters, "and seats be built in tiers. At the very least, the place should seat twenty thousand people." Scopes's alleged crime bore the same penalty as the most minor liquor-law violations that Raulston heard in batches every day, but he set no such time limits on this trial. Bryan and Darrow would give speeches, he predicted. "In my estimation the trial is of such intellectual interest and importance that I believe it fair to give both sides ample time to present their cases," Raulston observed. "Big issues are involved."[61]

Scopes and Neal waited all morning on the courthouse lawn for the grand jury to hand down its indictment. When it came, Scopes told reporters, "I am ready to go through with it." Prodded for further comment, one journalist noted, Scopes "started to explain his theory that Darwinian evolution is not incompatible with the idea of an ever present, loving father of the universe. But [Neal] stopped him and led him down the street." One listening juror remarked, "Young Whippersnapper—ought to be hanged on the spot."[62] Scopes promptly left town for a visit with college friends and family in Kentucky. Neal returned to Knoxville. Stewart and Raulston resumed their regular rounds traveling the judicial circuit. The parties spent the next six weeks preparing for what some pundits heralded as The Trial of the Century, a shopworn designation borrowed from the previous year's Leopold–Loeb case. Anticipation mounted across the land.

— CHAPTER FIVE —

JOCKEYING FOR POSITION

THE INDICTMENT of John Scopes inflamed the righteous ire of Columbia University's influential president, Nicholas Murray Butler. "The Legislature and the Governor of Tennessee have with every appearance of equanimity just now joined in violently affronting the popular intelligence and have made it impossible for a scholar to be a teacher in that State without becoming at the same time a law-breaker," he told June graduates at Columbia's commencement. "The notion that a majority must have its way, whether in matters of opinion or in matters of personal conduct, is as pestilent and anti-democratic a notion as can possibly be conceived." Denouncing fundamentalists and antievolutionists as the "new barbarians" storming the citadels of learning, Butler proclaimed, "Courage is the only weapon left by which the true liberal can wage war upon all these reactionary and leveling movements. Unless he can stand his ground and make his voice heard and his opinions felt, it will certainly be some time before civilization can resume its interrupted progress."[1]

Butler's widely reported address served as a clarion call for educators and helped set the tone for liberal reaction to the upcoming trial. Fundamentalists represented a clear and present danger to progress that should be exposed and opposed in Dayton. Princeton president John

Greer Hibben denounced the antievolution law as "outrageous" and the Scopes trial as "absurd." Yale president James Rowland Angell expressed a similar viewpoint to the graduating class of his university, admonishing them that "the educated man must recognize and knit into his view of life the undeniable physical basis of the world."[2] The ACLU invited twenty prominent progressive educators to serve on a Tennessee Evolution Case Fund advisory committee to help raise money for the defense. All twenty accepted the invitation, including the two senior statesmen of American higher education, president emeritus Charles W. Eliot of Harvard and president emeritus David Starr Jordan of Stanford. From England, George Bernard Shaw condemned the "monstrous nonsense of Fundamentalism" and Arthur Keith confessed that he would like to see the Scopes "prosecution hanged on the spot." When asked for his opinion of the Tennessee law, Albert Einstein replied, "Any restriction of academic freedom heaps coals of shame upon the community."[3]

When Scopes visited New York in early June to confer with ACLU officials, the fashionable Civic Club hosted a formal dinner in his honor. The event attracted an overflow crowd, which the *New York Times* characterized as "comprising almost every shade of liberal and radical opinion."[4] Commenting on Scopes's visit, a reporter for the *New York World* observed, "Under the banner of liberalism, to the blaze of page headlines, with the aid of special interviews, posed photographs and human interest incidents, the conglomerate host of liberals is falling in about the lanky, grave-eyed Tennessee high-school teacher." According to this reporter, the "army" surrounding Scopes included "feminists, birth-control advocates, agnostics, atheists, free thinkers, free lovers, socialists, communists, syndicalists, biologists, psychoanalysts, educators, preachers, lawyers, professional liberals, and many others, including just talkers."[5]

The ACLU endeavored to project an all-American image for Scopes during this visit. Except for the Civic Club dinner, which the ACLU needed to raise funds for the defense, Scopes remained mostly in his hotel room or private meetings. At public appearances and press interviews, he simply repeated his account of the trial's humble origins: "Just a drug store discussion that got past control," as he put it. Hounded by the New York press for a juicier news item and influenced by the people around him, Scopes occasionally revealed a more radical side. "I don't know what the term 'parlor Socialist' means exactly, but I think that is what I am," he confided to one reporter. When asked by another if he was a Christian, Scopes let slip, "I don't know, who does?"[6] Nevertheless, carefully staged press photographs near the Statue of Liberty in New York and on Capitol Hill during a return-trip stopover in Washington communicated the desired image to millions. "John Thomas Scopes," the *Washington Post* reported, "yesterday stood before the original copy of the Constitution of the United States at the Congressional library and gazed long and wonderingly at the script upon which he hopes to vindicate his right to instruct the young of the land in accepted scientific principles." Here he returned to the script. "I want to correct the impression that I hold nonreligious views and am in fact an agnostic," he told the *Post* reporter. "While I am not a member of any church I have always had a deep religious feeling."[7]

To demonstrate scientific support for the cause, Scopes made public appearances in New York with three of America's best-known evolutionary scientists: the paleontologist Henry Fairfield Osborn, the psychologist J. McKeen Cattell, and the eugenicist Charles B. Davenport. All three men helped shape the public response to the upcoming trial.

Osborn had been wrangling with Bryan over the teaching of evolution for years and redoubled his efforts to promote the theory of human evolution and his views about its compatibility with Christian concepts of morality. "The facts in this great case are that William Jennings Bryan is the man on trial," Osborn argued in a hastily compiled book dedicated to Scopes and published on the eve of trial. Science gave irrefutable evidence of human evolution, according to Osborn, and "evolution by no means takes God out the Universe, as Mr. Bryan supposes." In this book, Osborn characterized the Commoner as blinded by "religious fanaticism" and "stone-deaf" to scientific argument—"he alone by his own resounding voice drowns the eternal speech of Nature." This became a common image of Bryan in the popular press: a typical political cartoon pictures Bryan's gaping mouth overshadowing a quiet researcher above the caption, SCIENCE AND SHOWMANSHIP. "Bryan's gospel is not truth," Osborn maintained, "it is an ill-starred state of opinion, disastrous to true religion, disastrous to morals, disastrous to education."[8] Clinging to an outmoded view of purposeful evolution that Darrow repudiated, Osborn publicly warned Scopes against "radicalism" at the trial and declined to appear as an expert witness for the Chicago agnostic.[9] Nevertheless, he posed for pictures with the Tennessee teacher among prehistoric fossils at his American Museum of Natural History and kept up a steady drumbeat against Bryan in magazine and newspaper articles during June and early July.

In his role as leader of the American Association for the Advancement of Science and owner-editor of its journal, *Science*, the efforts of Cattell on Scopes's behalf were less visible than those of Osborn but no

Editorial cartoon during the Scopes trial ridiculing Bryan's attacks on science. (Copyright © 1925 *New York World.* Used with permission of The E. W. Scripps Company)

less important. He met Scopes in New York to reaffirm the AAAS's commitment to the defense. "The American Civil Liberties Union can count on the association providing scientific expert advisors in defense of Professor Scopes," the AAAS promised in accord with a formal resolution drafted by Osborn, Davenport, and the Princeton biologist Edwin G. Conklin. The resolution declared that "the evidences in favor of the evolution of man are sufficient to convince every scientist of note in the world" and hailed Darwin's theory as "one of the most potent of the great influences for good that have thus far entered into human experience."[10] The resolution served as a launching pad for Cattell's efforts. As a scientist and AAAS officer, he worked closely with Maynard Shipley's Science League of America in orchestrating scientific support for the Scopes defense; as a science editor and publisher, he promoted the cause through editorials in *Science* and work with Watson Davis's Science Service, an agency that distributed popular articles about science to magazines and newspapers.

Davenport's involvement in the Scopes case began with a Science Service article entitled "Evidences for Evolution" that appeared in scores of newspapers across the country as the first in a series of pretrial columns in which prominent scientists capitalized on popular interest aroused by the case to educate the public about evolution. Davenport represented a logical choice for writing the initial article because, as America's lead eugenicist, he had a vital stake in defending the teaching of evolution. The textbook used by Scopes, Hunter's *Civic Biology*, featured Davenport's research into the evolutionary improvement of humans "by applying to them the laws of selection," and stressed the importance of proper "mate selection" in this process.[11] Eugenic mate selection required education, however, and Bryan had targeted eugenic thinking as one of the evil consequences of teaching evolution. Davenport struck back first in his Science Service article and later by giving his public blessing to Scopes in New York. "Fundamentalists accept what they have been told about the accuracy of description of the origin of the universe given in Scripture," Davenport noted in the Science Service article. "The biologist has his own idea of what is the word of God. He believes it to be the testimony of nature." He offered the laboratory breeding of new "forms of banana fly" as his evidence for evolution.[12] Later articles in the series featured other elementary evidences for evolution, such as the human tailbone and cultural development. With the Scopes trial only weeks away, such shopworn scientific evidence became newsworthy.

Throughout the country, scientists and educators reported widespread curiosity about the theory of evolution. Books on the topic sold briskly even in Tennessee, where Vanderbilt University chancellor James H. Kirkland predicted that the trial would stimulate "far more inquiry" into Darwinism. At the time, most Americans simply understood the theory of human evolution to mean that people came from apes. Bryan played on this common understanding in his public addresses, often repeating the popular applause line, "How can teachers tell students that they came from monkeys and not expect them to act like monkeys?" Osborn took this opportunity to explain otherwise. "The entire monkey-ape theory of human descent, which Bryan and his followers are attacking, is pure fiction, set up as a scarecrow," he commented in the *New York Times.* "Man has a long and independent line of family ascent of his own." Following Osborn's lead, Scopes made similar comments to the press. Referring to the massive outpouring of information about evolution, the world-famous horticulturist Luther Burbank described the trial as "a great joke, but one which will educate the public and thus reduce the number of bigots."[13]

Liberal ministers joined in the public outcry over the indictment and trial of John Scopes. These events unfolded at the height of the fundamentalist–modernist controversy, when intradenominational battles between liberal and conservative Christians made front-page headlines in newspapers across the country. The antievolution movement split those factions along the crucial fault line of an evolutionary versus a literal interpretation of the Bible. Neither side could afford to back down on the issue. Bryan stood as the recognized leader of fundamentalist forces within the northern Presbyterian church. His entry into the Scopes case brought a predictable response. "Practically every preacher in New York touched on the subject last Sunday," one newspaper reported the following week. "Many defended the law, but many others ridiculed it and scored William Jennings Bryan for his dramatic entrance into the case."[14]

Charles Francis Potter, a prominent Unitarian minister, led the assault on Bryan in Gotham that Sunday and hounded the Commoner all the way to Dayton. "Mr. Bryan beclouds the issue, saying that the choice is between fundamentalism and atheism," Potter declared. "If the voice which demands absolute acceptance of every word of the Bible is the only one to be heard, the less educated will begin to believe that voice." Potter and other modernists sought to provide other voices at this seemingly critical juncture. Indeed, warning that "we are just beginning to hear the Fundamentalist advance—the Tennessee trial is the opening barrage," he called on modernists to "take ten of the hundred reasons for doubting the Bible's literal truth and drop them from airplanes if necessary in cities of the South." As public interest in the Scopes trial mounted, Potter grew increasingly optimistic about its educational value.[15] "If the Anti-Evolutionists in Tennessee were aware of the existence of any other religions than their own, they might realize that it is the very genius of religion itself to evolve from primary forms to higher forms," he commented in a later sermon. "The author of the anti-evolution bill is obviously nearer in mental development to the nomads of early biblical times than he is to the intelligence of the young man who is under trial."[16]

Seizing the opportunity to educate and enlighten, Potter carried his message to the people of Dayton, originally with the expectation of serving as an expert witness on religion for the defense but ultimately as a freelance writer and speaker after he refused to endorse the defense's public position that the theory of evolution was compatible with scripture. Mencken captured the scene in a report from Dayton: "There is a Unitarian clergyman here from New York trying desperately to horn into the trial," Mencken observed. "He will fail. If Darrow ventured to put him on the stand the whole audience, led by the jury, would leap out of the courthouse windows, and take to the hills."[17] For Potter, the Bible reflected an earlier religious consciousness from which only Christ's moral teaching should be retained—and those should be integrated into an evolutionary world view. Rappleyea secured an invitation for Potter to preach this gospel in Dayton's northern Methodist church one Sunday during the trial, but opposition within that relatively liberal congregation to the so-called New York infidel forced a cancellation. He settled for delivering a Sunday evening sermon on the courthouse lawn. Despite Mencken's prediction, Potter also managed to give the opening prayer at trial one day—without any leapers—and took that occasion to petition for "the progress of mankind toward thy truth" from "Thou to Whom all pray and for Whom are many names." All other courtroom prayers were directed exclusively to the Christian God.[18]

Potter's place as a defense expert passed to the preeminent voice of modernism among American Christians, Shailer Mathews of the University of Chicago divinity school. Mathews could easily reconcile evolutionary science with the Bible through a modernist interpretation of scripture. "The writers of the Bible used the language, conceptions and science of the times in which they lived. We trust and follow their religious insight with no need of accepting their views on nature," he explained in a widely reported address delivered in Chicago shortly before he was to leave for Dayton. "We have to live in the universe science gives us. A theology that is contrary to reality must be abandoned or improved." This left Mathews between Potter and the fundamentalists: science informed scripture rather than the other way around, as the fundamentalists believed; but the Bible remained divine, which Potter denied. "He who understands the Bible in accordance with actual facts has no difficulty in realizing the truth of its testimony that God is in the processes which have produced and sustain mankind," Mathews maintained. For him, evolution was divine creation, and human religious understanding developed over time. At trial, Mathews offered to explain how the Genesis account of creation symbolized an evolutionary process, "and how that process culminated in man possessed of both animal and divine elements."[19]

Modernist ministers and theologians pressed the assault against fundamentalists in countless churches throughout the country. "William Jennings Bryan thinks that God ceased speaking to man after the first chapter of Genesis was written," one New York Methodist pastor proclaimed. "To make belief in Genesis and belief in Christ stand or fall together is absurd. The two beliefs are on different levels," a Michigan Baptist minister insisted. "Evolution is not on trial; Tennessee is," a California Congregationalist preacher added. "And the judgment has already been given by the high court of public opinion. The people of Tennessee are the laughing stock of the world."[20] Suddenly, for a few weeks, ministers could grab headlines anywhere in the country simply by asserting that the theory of human evolution did not conflict with the Bible.

Tennessee's beleaguered modernists took up the cause with especial fervor. Several Tennessee clerics offered to serve as experts for the defense, two of whom were picked to testify. Still others preached sermons attacking Bryan and the prosecution. A large crowd turned out in

Knoxville for a mid-June debate between two local fundamentalist and modernist ministers. "Today, theology is called upon to adjust itself to scientific facts," the modernist minister maintained. "Christian theology adjusted itself to the Copernican theory, and to the facts of geology and for a majority of Christian scholars the adjustment to the facts of evolution has already been made."[21] In Tennessee, such scholarship centered in Vanderbilt University, a liberal Methodist school. When the threat of antievolution legislation first arose, the university hosted a major address by the famed New York modernist Harry Emerson Fosdick. After the law passed, its chancellor vowed to continue the teaching of evolution within his private institution. Now, with the Scopes trial looming, school officials turned the June commencement exercises into a defense of evolutionary science. "Christ did not come into the world to dictate to scientists what they should think," declared the baccalaureate speaker, according to whom the theory of evolution harmonized perfectly with scripture. Indeed, he claimed that the Bible depicted an evolutionary development of religious consciousness, and urged the church to "canonize" Darwin and other scientists "under a special head: Servants of the truth of God." The commencement speaker echoed these themes.[22]

Middle ground did exist between modernism and fundamentalism but gained little attention in the public debate surrounding the Scopes trial. Each viewpoint was internally consistent, but many Americans opted for a pragmatic compromise that left room for both traditional religion and modern science by maintaining that orthodox belief in the Bible does not preclude an allegorical interpretation of the creation account. "A man can be a Christian without taking every word of the Bible literally," one defense expert on theology offered to testify at the Scopes trial. "When St. Paul said: 'I am crucified with Christ,' and when David said, 'The little hills skipped like rams,' neither expected that what he wrote would be taken literally." Similar textual interpretation allowed this witness to reconcile evolutionary science with the Genesis account by accepting evolution as God's means of creation. "I am thoroughly convinced that God created the heavens and the earth," he observed, "but I find nothing in the Scripture that tells me His method."[23]

Another Christian theology expert argued for the defense that science and religion could never conflict because they belonged to separate spheres of knowledge. "To science and not to the Bible must man look for the answers to the questions as to the process of man's creation," he offered to testify. "To the Bible and not science must men look for the answer to the causes of man's intelligence, his moral and spiritual being."[24] By presenting these two witnesses along with Mathews, the defense effectively demonstrated various ways that American Christians harmonized sincere religious faith with the findings of modern science.

The popular press seemed intent on pitting fundamentalists such as

Editorial cartoon suggesting that many Christians did not agree with Bryan's attacks on the theory of evolution. (Copyright © 1925 *New York World.* Used with permission of The E. W. Scripps Company)

Bryan and Riley against modernists such as Mathews and Fosdick, or against agnostics such as Darrow, all of whom scorned the middle. Bryan, for example, publicly dismissed theistic evolution as "an anaesthetic that deadens the Christian's pain while his religion is being removed," while Mathews rejected attempts to retain Mosaic concepts of morality without Mosaic concepts of creation.[25] During the twenties, these two extremes gained adherents at the expense of the middle—and each claimed to represent the future of Christianity. Their clash spawned the antievolution movement and well deserved the attention it received during the Scopes trial. Christians caught in the middle sat on the sidelines."The thing that we got from the trial of Scopes," a Memphis *Commercial Appeal* editorial observed, was that most "sincere believers in religion" simply wanted to avoid the origins dispute altogether. "Some have their religion, but they are afraid if they go out and mix in the fray they will lose it. Some are afraid they will be put to confusion. Some are in the position of believing, but fear they can not prove their belief," the editorialist noted, so they leave the field to extremists such as Darrow and Bryan.[26]

The middle did not remain entirely silent. President Hibbon of Princeton loudly complained about the trial, "I resent the attempt to force on me and you the choice between evolution and religion." Some religious scientists used the opportunity to promote nonmaterialistic theories of evolution. Many were modernists, like Osborn, but others were orthodox, such as the Vanderbilt University science professor who wrote into the *Nashville Banner*, "As a scientist, I believe that the theory of Lamarck concerning the inheritance of acquired characteristics is probably in the process of being verified."[27] That would resolve the controversy, the writer maintained. Such subtle arguments, however, attracted few headlines in newspapers bent on dramatizing the conflict between science and religion.

Even James Vance, the leading proponent of moderation within Tennessee's religious circles, added little to the public debate over the Scopes trial. Vance served as pastor of the nation's largest southern Presbyterian church and once held the denomination's top post. In 1925, readers of a leading religious journal voted him one of America's top twenty-five pulpit ministers, along with fundamentalist Billy Sunday and modernist Harry Emerson Fosdick. When the antievolution movement first began in 1923, Vance and forty other prominent Americans, including Conklin, Osborn, 1923 Nobel laureate Robert Millikan, and Herbert Hoover, tried to calm the waters with a joint statement that assigned science and religion to separate spheres of human understanding. This widely publicized document described the two activities as "distinct" rather than "antagonistic domains of thought," the former dealing with "the facts, laws and processes of nature" while the latter addressed "the consciences, ideals and the aspirations of mankind." It offered no reasoned reconciliation of the apparent conflicts between them, however.[28] In 1925, Vance joined thirteen moderate or liberal Nashville ministers in petitioning the Tennessee Senate to defeat the "unwise" antievolution bill. After reading the petition on the Senate floor, however, even an opponent of the legislation had to concede "that there was no reason assigned in the written request" for defeating the bill.[29]

Vance had plenty of company straddling the fence over the Scopes case. Most national politicians followed the lead of President Calvin Coolidge in dismissing the case as a Tennessee matter. Tennessee politicians tended to mimic their governor, who defended the law and denounced the trial, but who clearly wished to avoid the entire issue and vowed to stay away from Dayton. Very few state legislators attended the trial, despite offers of reserved seating. Even the law's author, J. W. Butler, only showed up after a newspaper syndicate offered to pay him for commentary on the proceedings. Labor unions hesitated to choose sides between two longtime friends—Bryan and Darrow. Only a handful of small unions did so, such as the prodefense Georgia Federation of Labor. Even the nation's two leading teachers' organizations split. Pushed by its vice president, ACLU executive committee member Henry Linville, the smaller American Federation of Teachers adopted a resolution in support of Scopes. The larger National Education Association rejected a similar resolution as "inadvisable."[30] Relatively little comment about the trial survives from African Americans. A few black evangelists, such as Virginia's John Jasper, endorsed Bryan's position, while the NAACP, which worked regularly with the ACLU, participated in some of the ACLU's New York meetings on the trial. In any event, the outcome would not affect African Americans, because Tennessee public schools enforced strict racial segregation and offered little to black students beyond elementary instruction.

White fundamentalists rushed in to fill the void, and willingly engaged modernists and evolutionists in setting the terms for public debate over the trial. In pulpits across America, conservative ministers argued against Darwinism. Many attacked Darrow and the menace of materialism as well, such as the Tennessee pastor who claimed that he "had been searching literature and the pages of history in an effort to find someone with whom he might class Darrow, but as yet had not been able to place him but in one class, and that of the Devil."[31] Leading antievolution crusaders such as Riley, Norris, Straton, Martin, and Sunday redoubled their efforts in the days before the trial, barnstorming the country for creationism. On a train to Seattle, Norris wrote to Bryan, "It is the greatest opportunity ever presented to educate the public, and will accomplish more than ten years campaigning."[32] From Oregon, Sunday added his endorsement of "any views expressed by William Jennings Bryan."[33] Summer having come, the Bible conference and Chautauqua seasons were in full swing, providing ready audiences of antievolutionists.

During the twenties, the public became fascinated by formal debates between proponents and opponents of the theory and teaching of evolution. In 1924, for example, Straton and Potter clashed over the theory before a large audience at Carnegie Hall in a debate broadcast live on the radio and subsequently published by a commercial press. A panel of three judges from the New York State Supreme Court gave a unanimous decision to Straton on technical merit. "With the exception of the legal battles to outlaw evolution or to get 'scientific creationism' into the public schools," the historian Ronald Numbers observed, "nothing brought more attention to creationists than their debates with prominent evolutionists."[34] Public interest in the coming trial generated a variety of such debates across the country, including a series between Riley and the science popularizer Maynard Shipley on the West Coast. "Please report my compliments to Dr. Riley," Bryan wrote to Straton in early June, just before Straton joined Riley for the final debate in Seattle. "He seemed to have the audience overwhelmingly with him in Los Angeles, Oakland and Portland. This is very encouraging; it shows that the ape-man hypothesis is not very strong outside the colleges and the pulpits."[35] For the moment, at least, antievolutionists appeared to have the upper hand.

The presence of Riley or Straton insured a large audience, but a pair of mid-June debates in San Francisco between Shipley and two young editors of a Seventh Day Adventist journal may have attracted the greatest attention. According to the *San Francisco Examiner*, "That the Scopes trial is a living issue in San Francisco as elsewhere was indicated by the large crowd which on both evenings filled the auditorium long before the meeting hour, and afterwards filled the street and threatened to rush the doors."[36] Prominent California jurists served on the panel of judges. Shipley spent the first debate sniping at Bryan, which allowed his Adventist opponent to win a split decision against the proposition, "Resolved, That the earth and all life upon it are the result of evolution," by systematically raising a host of technical questions about that theory. The second debate focused on the timely issue of teaching evolution in public schools. Here Shipley gained the victory with a plea for freedom. In typical Adventist fashion, his opponent presented the teaching of evolution as "subversive of religious views" and argued for "neutrality on the questions of religion" in public schools. The remedy: "Keep evolution and Genesis both out." Shipley countered with stories about the religious persecution of Galileo and Columbus for their scientific theories, and asserted the near universal support among scientists for the theory of evolution. "We hold that this theory, or any theory, advanced by those best qualified by education and experience to judge such matters, should be made known to the pupils of our publicly supported educational institutions, and that to suppress such knowledge is a social crime."[37]

The results of the San Francisco debates suggested that, in the spirit of liberty, people who doubted the theory of evolution might still tolerate the teaching of evolution. Perhaps Bryan sensed this all along and only campaigned to prohibit the teaching of evolution *as true;* but now he had to defend a broader law that barred all teaching about human evolution, while the defense followed Shipley's approach by pleading for individual liberty to learn and teach about scientific theories.

Despite strenuous efforts to reach the public through debates and addresses, Scopes's opponents regularly complained that the press garbled their message—reflecting in part their own perceptions. Following the San Francisco debates, for example, the Adventist science educator George McCready Price wrote to Bryan, "Our side whipped Mr. Shipley 'to the frazzle.' " Yet newspaper reports were mixed, as were neutral judges' and audience reactions; even an accurate news account of antievolution arguments might not sound as good as proponents remembered them. Accordingly, Price directed Bryan to "the full report of the debate" as published by an Adventist press.[38] In a private letter written shortly before the Scopes trial, Bryan explained his criticism of the press regarding the antievolution controversy. "I think the newspapers desire to be truthful about matters of science. Whether they are thoroughly sensible depends a good deal upon one's point of view," he commented. "I do not consider it thoroughly sensible for a paper to publish as if true every wild guess made by a man who calls himself a scientist; and yet the wilder the guess the more likely it is to be published."[39]

In fact, some bias against the prosecution did taint the news coverage. Most major American newspapers went on record favoring the defense. Even within Tennessee—although editorialists roundly criticized Dayton for staging the trial and several of them grudgingly conceded that the court should enforce the law—only one major daily newspaper, the Memphis *Commercial Appeal,* consistently supported the prosecution. Surveying the initial press commentary, a *Nashville Banner* editorialist observed that "There are vigorous champions of the right of the state to regulate its institutions, but a great many editors commenting insist that the question is whether truth shall be limited by law. Inevitably Mr. Bryan has become something of the storm center." During the trial, an article in a trade publication for journalists commented, "Some of the reporters are writing controversial matter, arguing the case, asserting that civilization is on trial. The average news writer is trying to stick to the facts as revealed in court, but it is a slippery, tricky job at best."[40] Based on a later study of editorial and news articles from the period, the journalism professor Edward Caudill agreed: "The press was biased in favor of Darrow," but mostly due to its insensitivity to faith-based arguments rather than to intentional advocacy.[41]

Whatever the source for bias, the results could be quite blatant. For example, when T. T. Martin passed through Chattanooga on his way to the trial he defended antievolution laws with the standard claim that they protected the individual liberty of religious students. Apparently unable to see any connection between the restrictive statute and individual liberty, the *Chattanooga Times* article on Martin's speech dismissed his claim as "quite novel."[42] Stung by critical letters to the editor from fundamentalists, the newspaper's managing editor sought a balance by commissioning Chattanooga's leading fundamentalist minister to join the paper's regular staff reporters in covering the trial "with no restrictions," as the minister was told, "save the truth and nothing but the truth be written." This policy, which Bryan hailed as "highly commendable," produced a diverse array of articles, with the minister's daily features typically published alongside those written by modernist clerics or Watson Davis's Science Service.[43] No other newspapers followed this approach, however. In mid June, when Riley, Martin, and other prominent antievolutionists offered a series of newspaper columns to balance the proevolution Science Service series, there were few takers among major papers.

Antievolutionists despaired of receiving fair treatment in the secular press. A letter to Sue Hicks from his brother Ira, a fundamentalist pastor in New Jersey, captured this feeling of frustration. "I have no doubt about the outcome of the case," Ira wrote in mid June. "What I fear is the news papers will color everything to look like a victory for evolution as their sympathy is there. To get the real facts of this case before the people, especially in the north, is going to be a difficult task."[44] Alternate outlets for information existed in church newspapers and journals. Some supported the prosecution, such as *The Baptist and Reflector*, which sent its editor from Kentucky to Dayton to cover the trial. Another Baptist journal offered its support from afar: "Scopes is just a fool boy who has lent himself to be the tool of faddists and opportunists."[45] A pretrial article in a Washington, D.C.-based fundamentalist journal, *The Present Truth*, added, "Scopes as a teacher is an employee of the State, paid out of state funds, and surely the State has a right to say what he may do and may not do in his official capacity."[46]

Most traditional church publications appeared under denominational auspices, however, and many established denominations were split by the fundamentalist–modernist controversy, which left their newspapers and journals in the middle on the Scopes case. Some criticized both Bryan's fanaticism and Darrow's naturalism; others called for tolerance or simply avoided the issue. In discussing the trial, Roman Catholic newspapers warned parishioners against both the theory of evolution as materialistic dogma and antievolution laws as part of an effort by Protestant fundamentalists to control public education. The Catholic Press Association sent a top officer, Benedict Elder, to cover the trial for diocesan newspapers across the country. Upon his arrival in Dayton, Elder complained about the "religious complex [of] some writers for the metropolitan papers," and offered his qualified support for the prosecution: "Although as Catholics we do not go quite as far as Mr. Bryan on the Bible, we do want it preserved."[47] Elder went to Dayton with a top Knights of Columbus official. "There is a vast amount of sympathy for Mr. Bryan and the state of Tennessee among the Catholics of America," the official noted. "However one may differ from him, the efforts of the Great Commoner serve the Christian faith of the young of Tennessee, and he is entitled to respect."[48]

Antievolutionists increasingly turned to interdenominational journals and publishers to communicate their side of the story. The WCFA's quarterly journal presented its view of the Scopes trial to the faithful, and America's two leading conservative Christian magazines, *Moody Monthly* and *Sunday School Times*, also took up the cause. Fundamentalist publishing houses, particularly the nondenominational Fleming H. Revell Company, contributed to the barrage of words. Antievolution books by Bryan sold so well that he discussed retiring from the lecture circuit after the Scopes trial to concentrate solely on writing. T. T. Martin's *Hell and the High School* and Price's *The Phantom of Organic Evolution* chalked up record sales; indeed, Martin hawked his book near the courthouse in Dayton under a large banner bearing the book's title, which created a popular backdrop for photographers who wanted to emphasize the trial's carnival atmosphere. The role of interdenominational and parachurch organizations in American religion had been increasing for years as traditional churches divided into liberal and conservative factions that crossed denominational lines; events leading up to the Scopes trial, however, accelerated this trend—especially for fundamentalists. Just before the trial, for example, when Riley announced the formation of a half dozen local societies to push for antievolution laws in various states, he stressed, "The societies are sponsored by fundamentalists of all denominations."[49]

Bryan moved at the center of the fundamentalists' pretrial publicity campaign. He kept in close contact with leading antievolutionists as they spoke around the country. He traveled extensively himself, crisscrossing the eastern United States half a dozen times during May and June, speaking freely about the case in a style reminiscent of his whistle-stop campaigns for the presidency. The trial "is not a joke," Bryan assured a Chicago audience, "but the beginning of the end of attacks upon the Bible by those teachers in the public schools who have been substituting the guesses of scientists for the word of God."[50] Before a crowd of over 20,000 people in a small midwestern town, he added, "The most important elements that stir the human heart are bound up in [this case:] the education of the child and the religion of the child."[51] In full campaign mode, the Commoner proclaimed in Brooklyn, "We must win if the world is to be saved."[52] Back home in early July, he reported to the Miami Rotary Club: "The wide publicity given evolution and religion is focusing the attention of the world on a subject the people did not fully understand."[53] Upon meeting Scopes in Dayton several days later, Bryan leaned toward the teacher and quietly said, "You have no idea what a black and brutal thing this evolution is."[54]

Bryan's busy schedule made it difficult for the prosecutors to arrange a joint strategy session. Knowing that Bryan was passing through Tennessee in early June, Sue Hicks proposed that the Commoner stop over in Chattanooga for a conference, but Bryan had a speech in Tallahassee the following day. "You might meet me in Nashville at 8 A.M., [and] ride to Decatur," Bryan scribbled his reply on hotel stationery. "This would give us about four hours together on the train, which would I think be sufficient for plans necessary now."[55] Thus forewarned of Bryan's itinerary, a band playing "Onward Christian Soldiers" and a blue-ribbon delegation of city and state leaders greeted the Commoner's train when it pulled into Nashville. The three Dayton prosecutors, Hicks, Hicks, and Haggard, met with Bryan that morning; Stewart was in court at the time. The prosecution met together only once more prior to assembling in Dayton, late in June when Bryan had a brief stopover in Atlanta. Otherwise, they communicated by mail. Nevertheless, a bond immediately formed among the prosecutors. Four days after the first meeting, Sue Hicks wrote to his brother Ira, "We had a splendid conference with Bryan . . . in Nashville and rode with him in [his] state room to Chattanooga. He is greatly enthused about the case and will talk about nothing else. Of course we think Bryan is a wonderful man."[56] In similar letter to another brother, Sue Hicks added that Bryan "is making great plans for this case. He says it is [a] turning point for Christianity."[57]

Bryan never varied in his public pronouncements regarding the prosecution's strategy. "I have been explaining this case to audiences. It is the *easiest* case to explain I have ever found," he wrote to Sue Hicks at the outset. "The *right* of the *people* speaking through the legislature, to control the schools which they *create* and *support* is the real issue as I see it." Bryan went on to add, "By the way I don't think we should insist on more than the *minimum* fine and I will let the defendant have the money."[58] He reasserted this position after consulting with co-counsel on the train in Nashville. "The New York papers have entirely mistaken the issue," he told reporters. "Mr. Scopes demands pay for teaching what the state does not want taught and demands that the state furnish him with an audience of children to which he can talk and say things contrary to law. No court has ever upheld any such proposition." As to raising "the question of evolution" at trial, Bryan commented, "I am not so sure that it is involved."[59]

Privately, however, Bryan hoped to discredit the theory of evolution through expert testimony. Sue Hicks explained the plan to his brothers shortly after the prosecution's first strategy session. "We can confine the case to the right of the legislature to control the schools and easily win. However we want both legal and moral victory if possible," he wrote in strict confidence. "After we have put on sufficient proof to show the facts of the teaching, the state will rest its case and wait for the defense to move. They will likely want to win a moral victory for their scientific beliefs and will introduce various scientists, to substantiate the theory of evolution." Here, Bryan hoped to ambush the defense. "We are planning to meet them on every issue raised and we think, without trouble, we have them beat in both the legal and scientific phases," Hicks boasted. "It is part of our plan to keep the defense thinking that we are going to restrict the case to the right of the legislature to control, but when the trial comes on we can gain a moral victory by opening out the field to our evidence."[60] Bryan confided his hopes in a letter to Johns Hopkins medical school professor Howard A. Kelly, one of the scientific experts solicited to testify. "The American people do not know what a menace evolution is—I am expecting a tremendous reaction as a result of the information which will go out from Dayton, and I am counting on you as one of the most powerful factors," he wrote.[61]

Early on, the prosecution divided up responsibility for preparing and presenting the case. Recognizing his lack of trial experience and unfamiliarity with Tennessee law, Bryan left the legal issues strictly to the local attorneys. He assumed responsibility for securing scientists and theologians to testify against the theory of evolution. It was here that Bryan's ambitious plans for attacking the theory at trial began to break down. None of the Tennessee prosecutors knew anything about science. Sue Hicks's confidence about winning the scientific phase of the trial rested solely on Bryan's assertions about the matter. "Mr. Bryan is getting up the witnesses for us," he wrote to his brother Ira, "and expects to have many of the leading scientists and doctors of divinity."[62] This great expectation met with bitter disappointment.

During the first strategy session, Bryan referred to the work of George McCready Price in refuting the theory of evolution. Sue Hicks also heard about Price from his brother Ira, who called Price "one of the best geologists."[63] But Price carried no authority as a scientist outside fundamentalist circles. He lacked formal scientific training and devised his idiosyncratic geological theories about a recent six-day creation and cataclysmic Noachian Flood based on a literal reading of scripture informed by writings of the Adventist prophet Ellen G. White. Adventism stood on the fringes of fundamentalism, however, and Price's work gained only qualified support from Bryan and other prominent antievolution crusaders of the twenties—many of whom accepted a long geologic history of the earth based on a "day/age theory" or "gap theory" interpretation of the Genesis account. Prosecutors turned to Price as their principal scientific expert against the theory of evolution. "You are one of the outstanding scientists who reject evolution as a proven hypothesis," Bryan wrote to Price in early June. "Please let us know at once whether you can come."[64] But Price was lecturing in England and unable to return. "I do not think that I could do any good, even if I were present at the coming trial," Price wrote in a letter to Bryan. "It seems to me that in this case, it is not a time to argue about the scientific or unscientific character of evolution theory, but to show its utterly divisive and 'sectarian' character, and its essentially anti-Christian implications and tendencies. This you are very capable of doing."[65]

No other potential scientific expert contacted by Bryan wanted to participate. Several turned him down flatly. Only Kelly gave a qualified yes, writing that "the Christian must stand very literally with the Word regarding the creation of man," but he acknowledged "a possible continuous sequence in the life history of the lower creation."[66] In other words, nonhuman species evolved. This troubled Bryan from a strategic standpoint. "I would not be concerned about the truth or falsity of evolution before man but for the fact that a concession as to the truth of evolution furnishes our opponents with an argument which they are quick to use," Bryan wrote back. "If we concede evolution up to man, we have only the Bible to support us in the contention that evolution stops before it reaches man." Of course, this was Kelly's point when he offered to stand with the word of God rather than the evidence of science regarding human evolution. The prosecution had plenty of potential religious experts with better theological credentials than Kelly (Riley, Straton, and Norris offered to testify), so Bryan put Kelly on standby status. "I don't want to put you to the trouble of going to Dayton unless it is necessary," Bryan wrote, and it would not be necessary if the court foreclosed all scientific testimony; this became the prosecution's single-minded objective by the time of trial.[67]

As the trial date approached, Bryan began to worry about the composition of the prosecution team. He had joined the prosecution when questions still existed as to whether anyone in Dayton seriously wanted to enforce the law. Those doubts should have ended when circuit attorney general Tom Stewart took charge of the prosecution, even though the case only involved the type of misdemeanor typically left to city or county attorneys. Bryan dealt mainly with the original Dayton lawyers, all of whom lacked trial experience—as did the Commoner. The defense had assembled four of the finest trial attorneys in America, and Bryan was concerned. "While I think you and your brother, Mr. Haggard, and myself might be able to meet their attack without any outside help," he wrote to Sue Hicks in mid June, "I feel that the case is so important that we should not take any chances." Bryan went on to state that he had already informally asked "two prominent men from the outside to assist us so that our side will look as large as theirs.[68] The choice was revealing, and clearly would have achieved the stated objective by broadening the case beyond the issue of fundamentalism. One was Samuel Untermyer of New York, vice president of the American Jewish Congress; the other was Senator T. J. Walsh of Montana, a Roman Catholic.

In his letter to Hicks, Bryan described Untermyer as "the biggest lawyer I know," and the Commoner knew many of America's leading attorneys. Untermyer's father, a Jewish immigrant to Virginia, had fought for the Confederacy. After the Civil War, the boy went with his widowed mother to New York, where he rose to become a fabulously wealthy corporate lawyer and civil rights activist. Untermyer served as a leader in both the ACLU and the American Jewish Congress, but Bryan knew him from their work together for the Democratic party. "Being a Jew," Bryan wrote to Hicks, "he ought to be interested in defending Moses from the attacks of the Darrowites."[69] And Untermyer would have done so, except for the fact that he had just sailed for Europe. Bryan's letter caught up with Untermyer in London, from where he cabled detailed advice. Untermyer fully agreed that the legislature should control the school curriculum. "The most important question that will arise upon the trial, as I see it, is to restrain the defendants from reaching from outside the real issues of law that are involved in the controversy," he wrote. "I would seek to exclude all discussion by experts or otherwise on the subject of evolution. . . . If the Court is prompt and intelligent in its ruling the trial will be a rather perfunctory affair. If you and your associates would like to have my participation on the appeal I shall be glad to act."[70] This advice, coupled with the prosecution's problems in securing scientific experts, convinced Bryan to stick to a narrow legal strategy.

The local prosecutors disliked Bryan's idea of asking out-of-state attorneys to join the team. Except for locally popular figures such as Bryan, they argued, such attorneys carried little weight in Tennessee courts. "We some what doubt the advisability of having a Jew in the case," the Hicks brothers bluntly wrote to Bryan. Catholics posed a problem as well. Sue Hicks already had gloated to reporters over the prospect of besting the ACLU and Darrow. The former was "procommunist," he noted, and as to the latter, "All we have to do is to get the fact that Mr. Darrow is an atheist . . . across to the jury, and his case is lost." Now the brothers pleaded with Bryan, "We feel that it will be a great victory for our cause to whip them without additional counsel." They acknowledged their own inexperience, but stressed, "Attorney General Stewart is a good constitutional lawyer, a close observer, a good reasoner, a hard worker, and a good speaker. We feel that, under the conditions, he alone will be able to take care of [legal matters]." Bryan bowed to their objections, and left Stewart in charge.[71] Walsh quietly let the matter pass, while Untermyer, who wanted to honor his final pledge to Bryan, had to be told that his assistance was not wanted for the appeal. Two additional lawyers joined the prosecution team, however. The circuit's retired attorney general, Ben G. McKenzie, appeared alongside his son for the prosecution, and William Jennings

Bryan, Jr., then in private practice in California following service as a U.S. attorney in Arizona, arrived to help his father.

Bryan's penchant for oratory notwithstanding, the strategy and composition of the prosecution promised a quick trial. In a formal letter to the court submitting their witness list, the Hicks brothers wrote, "We have no list of witnesses to give out other than those we used before the grand jury. As we understand it, it is the duty of the Court to look within the four corners of the act and from that determine its constitutionality." They added a barb typical of remarks at the trial: "We have no desire to violate a rule of evidence and allow the defense to turn loose a slush of scientific imagination and guess work upon our people, upon whom from reports, these great lawyers from the north and northwest look with pity and compassion, denominating them a set of ignoramuses."[72] Governor Peay communicated similar advice to the court. In a public letter released shortly before trial, he declared, "The case should be tried in an hour. It is about as simple a proposition as could be stated and the great hurrah about it is unnecessary and unfortunate."[73] A prominent Nashville jurist felt the same way. "The question of whether or not the state has the right to prescribe a curriculum for its schools is the question upon which the Scopes trial should turn," he declared, "and if it does the trial will be a short one and rather uninteresting."[74]

Bryan and Stewart knew that Darrow and company would not quietly accept such narrow limits for the trial. "If we can shut out the expert testimony," Bryan predicted in a private letter shortly before trial, "we will be through in a short time. I have no doubt of our final victory, but I don't know how much we will have to go through before we reach the end."[75] Stewart anticipated a fierce fight. "The trial proper should be comparatively simple," he observed. "This challenges the right of the legislature to regulate the public schools in the state. . . . The legal questions, however, are about to be lost sight of in the consideration of this unusual matter."[76]

The defense, of course, took an expansive view of the "legal questions" raised by a state law against teaching evolution, and adopted a strategy calculated to push them to the fore. Darrow and Malone told the press that their case would take a month to present. Hays explained the reasons why. The Tennessee statute expressly outlawed teaching that denied the biblical account of creation. As a legal matter, according to Hays, this was "unconstitutional because, in the light of present-day knowledge of evolution, to be adduced from scientists, it is unreasonable." Further, he added, "the law was indefinite as well as unreasonable, because no two persons understand the Bible alike." Hays elaborated on this second proposition. "If the fight of liberalism and honest thinking is to be won it must have the support of millions of intelligent Christians who accept the Bible as a book of morals and inspiration," he explained. "Evidence which would tend to show that there is no conflict between religion and science, or even between the Bible, accepted as a book of morals, and science, would be more effective in answering their claims than a mere contention that the schools must be free to teach what these fundamentalists regard as irreligion." With Darrow, Malone, and Hays in control of the defense, the fight for individual liberty against majority control expanded to include scientific evidence for evolution and religious theories of biblical interpretation. "That the people should derive light and education from court proceedings may be novel," Hays wrote, "but it can hardly be objectionable."[77]

Defense attorneys began their efforts to enlighten and educate the public almost immediately through pretrial tactics that differed markedly from those of the prosecution. Although Bryan spoke widely about the menace of Darwinism, the prosecution kept as quiet as possible about their plans for the trial and said nothing in public about potential expert witnesses. The defense, in contrast, spoke openly about its plans and issued almost daily announcements about various scientists and theologians who would—or might (it was never quite clear)—testify on Scopes's behalf in Dayton. In late June, for example, Malone announced a list of ten distinguished scientists who "have already signified their willingness to serve as witnesses."[78] Only two of these ten actually went to Dayton, and the top names on the list—Osborn, Conklin, and AAAS president Michael I. Pupin—by this time clearly had said no. Defense attorneys suggested also that Luther Burbank would testify, even though the famed horticulturist had only agreed to serve on their advisory committee. The constant dribble of names insured a succession of newspaper articles linking the Scopes defense to America's most respected scientists, which helped to enlighten the public about the widespread support for the theory of evolution throughout the scientific community. Such a tactic also kept the prosecutors off balance, especially as their own well of scientific experts came up dry.

Bryan tried to dismiss defense experts with populist oratory, often decrying that a "scientific soviet is attempting to dictate what shall be taught in our schools." At trial, Bryan added, "It isn't proper to bring experts in here to try to defeat the purpose of the people of this state by trying to show that this thing that they denounce and outlaw is a beautiful thing."[79] He worried most about the public impact of Burbank's activities in support of teaching evolution and sought to discredit them. "I remember seeing a letter from [Burbank] which was published in Ohio in which he denounces religion," Bryan wrote to Riley three months before the trial. "Would it not be well for you to have some friends of yours in Minnesota write to him as if from the standpoint of an atheist and congratulate him on his activities and draw out from him a declaration of his atheistic views?" Bryan had good reason for concern. People everywhere know about Burbank's knack for breeding new commercial plant varieties, which seemed like an example of evolution at work. "There is no such thing as evolution," Bryan said in frustration shortly before trial, "Burbank? Ah, he merely produced varieties within a species."[80] Typical Rhea County jurors, however—most of whom were farmers—surely would listen attentively to Burbank. Indeed, when Hays introduced a statement from Burbank in court, Judge Raulston jerked up in his chair, "Is he here?" and was visibly disappointed to learn that Hays only offered a written statement.[81]

In all likelihood, the eight scientists who finally showed up for the defense were completely unknown to the people of Dayton. From the scene of the trial, the *Chicago Tribune* reporter Philip Kinsley attributed what he described as the defense's "trouble in getting prominent men to come here to testify on the side of evolution" to their fear of facing cross-examination by Bryan in a hostile setting. The best-known potential experts for the defense—Osborn, Davenport, Cattell, Burbank, Conklin, and David Starr Jordan—had clashed willingly with Bryan in public over evolution, however. They seemed more troubled about appearing at the trial with Darrow than against Bryan. Certainly none of them liked appearing in a supporting role opposite a showman such as Darrow. The Chicago attorney's radical agnosticism made some of them uncomfortable as well. Furthermore, all six championed coercive eugenic measures to guide human evolution, measures that Darrow denounced as incompatible with human rights. Hints of each of these reasons appear in the scientists' statements, but their absence from the trial spoke loudest of all.

This left the controversial Chicagoan as the only member of the defense team who could compete on the public stage with Bryan. Darrow did his best to promote Scopes's side of the dispute in a series of widely reported speeches and press statements during the month before trial. His late June visit to Dayton and Knoxville attracted the most attention. "The night he arrived there was a violent storm," Mencken joked from Dayton, "and horned cattle in the lowlands were afloat for hours."[82] Despite such efforts to sensationalize the contrast between Darrow and Daytonians, townspeople immediately took to the great agnostic. "He arrived wearing a straw katy, his coat open in a gesture of summer casualness," Scopes later recalled. "It was easy to like him. He drawled comfortably and hadn't any airs. He gave the impression he might have grown up in Dayton, just an unpolished, casual country lawyer, so ordinary did he act."[83] Darrow sized up the town, conferred with Scopes, and met the press. The Progressive Dayton Club hosted a banquet in his honor, which gave him a formal opportunity to explain his views on evolution and religion. "People of Dayton like his personality and think he is a great man," Sue Hicks reported to Bryan in a letter, "but they are all shaking their heads about his beliefs."[84]

Darrow was not speaking simply to Daytonians, however, but to all Americans. In speech after speech, he stressed the trial's significance. "This case is a difference of opinion of people upon a matter which effects life," he told the Progressive Club. "The country has fallen upon evil times. It seems that every organization has some law it is endorsing to force upon the people," he warned a large public audience at Neal's law school in Knoxville. "If the human race is going to be improved," Darrow asked in a New York address, "who will do it? The Bryans? . . . It is best to leave everyone free to work out things for himself. Nature is doing it in a big, broad way and doing it pretty successfully." From his naturalistic, materialistic perspective, Darrow cried out for tolerance and liberty: "What we are depends on heredity and environment, and we can control neither. As a result, I never condemn, never judge."[85]

His usual approach to a trial was quite different. "Ordinarily, Darrow's strategy was to dissipate the prejudice aroused by any crime of which the defendant might be accused . . . by good humor and light quips," Hays wrote, noting as an example Darrow's crack at a trial of a spouse killer, "Well, it was his own wife, wasn't it?"[86] Here Darrow sought the opposite effect so as to emphasize the threat to freedom and to counter Bryan's claim that an evolutionary world view offered no basis for morality. It worked. Hicks privately described Darrow's Progressive Club address as "wonderful." "Those who want to hear a great burst of oratory did not hear that," a deeply moved journalist wrote of Darrow's hour-long Knoxville lecture. "They simply saw a stooped man in baggy dark clothes who talked to them in ordinary conversational manner. They saw a tired but kindly face, shrewd eyes which often evoke laughter, but seldom laugh. And they liked it."[87] This contrasted starkly with Bryan and his bombastic majoritarian crusade for legal restrictions on academic freedom.

In what was scheduled as the highlight of his June visit to Tennessee, Darrow almost had the opportunity to present his side of the case at the annual meeting of the state bar association, but its president revoked the invitation when delegates became embroiled in controversy over the pending trial. Supporters of a floor resolution condemning Dayton for using a criminal trial as an "advertising medium" clashed with proponents of one demanding repeal of the antievolution statute, which delegate Robert S. Keebler of Memphis denounced as "half pitiful, half ludicrous" in an hour-long oration that systematically detailed constitutional objections to the law. As the meeting reeled toward chaos, the president ruled the whole topic out of order, struck Keebler's remarks from the record, and withdrew Darrow's invitation.[88] The ACLU subsequently printed two thousand copies of Keebler's "banned" oration, which it distributed in a bulk-mail solicitation for contributions to a special defense fund for the Scopes case. "The public's interest in the Scopes trial has been greater perhaps than any since the famous Dred Scott decision," the ACLU announced in launching this fund drive. "We believe that citizens all over the United States will want to have a part in this issue that will shape the future course of education in the country."[89]

Two of Darrow's co-counsels also spoke freely with the press and public during the weeks before trial, reinforcing Darrow's efforts to communicate the significance of the case. "No more serious invasion of the sacred principle of liberty than the recent act against the teaching of evolution in Tennessee has ever been attempted," Dudley Field Malone told a Knoxville women's group during his pretrial visit to the state.[90]

About the same time, John Neal warned a Chicago audience, "If the state's charges against Scopes are sustained you will see other evolution trials and perhaps a movement in congress to control the thought as well as the actions of people."[91] Following a pretrial visit to Dayton, Darrow's best-known co-counsel, Bainbridge Colby, issued a statement decrying the "holiday atmosphere surrounding the approaching trial," adding "the issue is grave in character, embodying principles at the base of our security of happiness and American citizenship."[92] The press resisted this characterization of events, and persisted in treating the entire episode as a joke. Editorial cartoons inevitably depicted the Great Commoner embroiled with monkeys—and the monkeys usually winning. Syndicated political humorist Will Rogers brushed aside an invitation to Dayton with the comment, "Bryan is due back here in the New York zoo in July."[93]

Malone assumed responsibility for pressing the defense contention that the theory of evolution did not conflict with the biblical account of creation. Taking this message to Baptist Tennessee during his pretrial visit presented a challenging role for a twice-married Roman Catholic divorce lawyer with Socialist ties, but Malone was a highly effective public speaker and the only professing Christian among the defense lawyers at Dayton. "I daresay that I am just as strong a believer in Christianity . . . as Mr. Bryan," Malone told a Chattanooga luncheon audience. "I find no difficulty in holding with devotion to Christianity and also to evolution. Theology is concerned with the aspiration of men and their faith in a future life. Science is concerned with the process of nature." These separate spheres need never cross, he added in an evening address to the local Civitan Club: "There should be no more conflict between religion and science than between the love a man gives to his mother and to his wife."[94] Prosecutors countered this message by attacking the messenger. "I read in today's Banner Mr. Sue Hicks' interview wherein he scored Darrow, Malone and other atheists," a Nashville legal advisor for the prosecution wrote to the Hicks brothers. "This is the line to attack, and you will find it most vulnerable and will strike the responsive chord with the people."[95]

Neal, for his part, kept the focus on academic freedom. He claimed to "represent the protest of the intellectuals of the south against the antievolution legislation," which he blamed on the "arrested [intellectual] development" of the region. "It is not a case of religion against irreligion, not a case of Fundamentalism against Modernism, but a case for the freedom of speech and thought," Neal told a New York audience.[96] Neal gained widespread attention in early June when he tried to delay the selection of new biology textbooks for Tennessee public schools until after the Scopes trial. His legal threats were ignored by the state textbook commission, which replaced Hunter's *Civic Biology* with texts that barely mentioned evolution.

Except for an occasional press release, the usually talkative Colby said little in public about the upcoming trial. He appeared ill at ease in Dayton during his pretrial visit and positively appalled when observing a criminal prosecution in Kingston, Tennessee, on the drive back to Knoxville. An accompanying reporter from the *Chattanooga Times* described it as a murder trial; Scopes recalled it as a rape case. By either account, a young defendant (of whom Scopes said, "At best he was a moron; more likely he was an imbecile") was being railroaded without proper representation in a courtroom filled with gun-toting spectators. Darrow had to be physically restrained from intervening, while Colby hung back moaning, "Those poor, poor unfortunate people." Upon his return to New York, he convinced the ACLU to seek an injunction to remove the Scopes case to a "sedate" federal court on the flimsy grounds that the antievolution statute applied to an institution receiving federal funds, namely, the University of Tennessee. After a federal judge abruptly denied this last-minute petition, a ruling even the other defense lawyers viewed as correct, Colby quietly resigned from the case. The *Chattanooga Times* reporter had predicted this development: "When he took one look at the hardy Tennessee mountaineers assembled in the Kingston courtroom, [Colby] departed in haste for the effete east, with the mental reservation that 'this is no place for me.' Colby saw at Kingston what he thought he would see at Dayton."[97] Darrow felt right at home, however, while Malone and the ACLU representative Arthur Garfield Hays approached the pending trial with a spirit of adventure.

Having survived a second attempt "to rob Dayton of the big show," as one reporter described it, townspeople eagerly completed preparations for the expected throng.[98] Officials roped off six blocks of the town's main street as a pedestrian mall, which quickly filled with hucksters and proselytizers. The state sent a mobile chlorination unit to provide an adequate supply of safe drinking water and a sanitary engineer to oversee waste disposal. Chattanooga contributed a fire brigade and six constables. A temporary tourist camp opened on vacant land owned by the coal company. Dayton's finest hotel, the Aqua, placed cots in its hallways, while the Ladies' Aid Society prepared to offer one-dollar lunches at a downtown church. Rappleyea fixed up an abandoned eighteen-room house known as the Mansion to accommodate visiting defense experts, leading some townspeople to joke that they used to think that the house was haunted, but now they knew it was. Mencken described it as "an ancient and empty house outside the town limits, now crudely furnished with iron cots, spittoons, playing cards and the other camp equipment of scientists."[99] The Morgan Springs Hotel, a nearby mountain resort, engaged a jazz orchestra to perform nightly during the trial. The local cinema screened *The She Devil.*

Dayton bustled with activity. Workers erected a speaker's platform on the courthouse lawn and marked off the county's first airstrip on a nearby pasture. Robinson's drugstore stocked up on books by both Bryan and Osborn, and hung out a banner declaring, "Where It Started." No need to define *It.* Other signs appeared along the main street, including several large banners proclaiming, READ YOUR BIBLE, one of which adorned the courthouse itself. A cavernous storage loft above a downtown hardware store became a makeshift press room for visiting reporters. Western Union stationed twenty-two key operators in town to transmit news reports and strung extra telegraph wires to nearby cities. The telephone company and post office hired additional staff. The Southern Railway added extra passenger service to Dayton and advertised free stopovers in town on all tourist tickets. The Progressive Dayton Club struck a souvenir coin bearing the likeness of a monkey wearing a straw hat.

The courtroom received a face-lift for the trial. A fresh coat of cream-colored paint brightened the walls. Five hundred additional spectator seats and a movie camera platform crowded the chamber. Telegraph wires ran into the courtroom for minute-by-minute transmissions of the proceedings, much like those used to broadcast big-league baseball games. The telephone company installed a bank of phones in an adjoining room, and new public toilets went in downstairs. In a move symbolic of the trial itself, the jury box was removed from the center of the chamber to make room for three central microphones, which fed loudspeakers on the courthouse lawn and in four

John Scopes, left, and his initial defenders, John Neal, middle, and George Rappleyea, right, beneath a trial-related poster on the way to a court hearing. (Courtesy of Bryan College Archives)

public auditoriums around town. WGN, the radio voice of the *Chicago Tribune*, arranged to transmit the message from the microphones through special telephone lines to Chicago, from where the station broadcast the proceedings live over the airways. "The event will be the first of its kind in the history of radio," the *Tribune* boasted, "undertaken as a demonstration of the public service of radio in communicating to the masses great news events." It dismissed concerns about the propriety of such a broadcast. "This is not a criminal trial, as that term is ordinarily understood," the announcement added. "It is more like the opening of a summer university. . . . The defendant, Scopes, is already a negligible factor. Nothing serious can happen to him. The contest is entirely over ideas."[100]

The composition of the town's population also changed. Many residents left Dayton during the trial, leasing their homes to visitors. The Bryan family, for example, occupied the modern home of a druggist, F. R. Rogers, who took his family to their cottage in the mountains. Darrow initially stayed in the Mansion but after his wife arrived from Chicago moved with her into the vacated home of a local banker. Malone checked into the Aqua Hotel with a striking woman who registered as Doris Stevens, which created quite a stir until townspeople realized that the woman (who registered under her own last name) was his wife. Hays and Neal spent most of their time at the Mansion, which served as headquarters for the defense team throughout the trial.

A diverse array of journalists, evolutionists, and antievolutionists trailed in behind the attorneys. Approximately two hundred reporters covered the trial for newspapers across America and as far away as London. Press photographers and newsreel camera crews also appeared in abundance. T. T. Martin preached in the streets, as did a Brooklyn rationalist who shouted about the evils of Christianity, and a Detroit man who billed himself as the Champion Bible Demonstrator. A small contingent of black Pentecostalists camped near town, attracting the attention of reporters who apparently thought that speaking in tongues was an indigenous Tennessee religious phenomenon, when in reality the great black church leader Charles Harrison Mason had brought Pentecostalism to the Tennessee African-American community from Los Angeles a decade earlier. "It sounds to the infidel like a series of college yells," Mencken wrote.[101] For a fee, anyone in Dayton could pose with a live chimpanzee or view fossilized remains of "the missing link."

Most visitors, however, had nothing in particular to say or sell but, as one African-American tourist from Atlanta told a *New York Times* reporter, simply "wanted to see the show." After surveying the crowd, the reporter concluded, "Whatever the deep significance of the trial, if it has any, there is no doubt that it has attracted some of the world's champion freaks."[102]

Bryan, the star attraction, arrived three days before the trial began. A summer heat wave pushed temperatures into the nineties that day and throughout the trial—twenty degrees above normal. While waiting for Bryan's noon train, a reporter asked a nearby bootblack, "Why all the crowd at the depot?" The reporter recorded the following response: "Des wait'n fur Willum Jennums Bryan, sir. . . . He's a hard-shell preacher, . . . a stand-patter, . . . a non-skidder, and de's no movin' uu'm when de thinks um right."[103] The Royal Palm limited from Miami finally arrived at 1:30 P.M. and made its first stop ever in the small town that it usually passed through at full throttle. "As Mr. Bryan stepped from the rear platform," one reporter observed, "he was greeted with applause and flutters of handkerchiefs. He was met by at least half the normal population of the town, and the temporary increase composed of newspaper people and photographers." Bryan wore a tropical pith helmet to protect his balding head from the sun and heat, and doffed it frequently to the crowd. "Just say that I am here," he declared with a broad smile. "I am going right to work, and I am ready for anything that is to be done."[104] This work consisted of a series of antievolution speeches around town.

Once the prosecution decided to oppose the admission of expert testimony at trial and narrowly limit the legal issue to majoritarian control over public education, out-of-court speeches and statements became the only sure way for Bryan to proclaim his message in Dayton. By arriving early, he now had the stage (and scores of news-hungry reporters) to himself. Bryan made the most of this opportunity. He strolled around town in his shirtsleeves greeting well-wishers and talking with reporters. He posed for pictures at Robinson's drugstore and lectured the school board on the dangers of teaching evolution. Bryan gave two public addresses before the trial began, one at a Progressive Dayton Club banquet in his honor and another in a dramatic mountaintop setting near the Morgan Springs Hotel. "The contest between evolution and Christianity is a duel to the death," Bryan said in explaining his view of the trial's significance to the Progressive Club. "The atheists, agnostics and all other opponents of Christianity understand the character of the struggle, hence their interest in this case. From this time forth Christians will understand the character of the struggle also."[105] On the mountaintop, he added, "Evolutionists, though admittedly in a minority, are intolerant enough to demand that the school teach their views, and their views really constitute their religion."[106]

The Commoner professed his faith in the judgment of the people, once the public was informed of the significance of the issue. "I have been quoted as saying that I think the decision of this case will be of importance," Bryan told reporters. "It is not the decision but the discussion which will follow that I consider important. It will bring the issue before the attention of the world."[107] No mere judicial decision could frustrate the awakened will of the people. "Who made the courts?" he asked in a rhetorical flourish before the Progressive Club. "The people. Who made the Constitution? The people. The people can change the Constitution and if necessary they can change the decisions of the court."[108]

Neal sat stone-faced throughout Bryan's Progressive Club address as those around him cheered. He stayed up late that night penning a formal response. "We regard Mr. Bryan's speech last night as the most remarkable utterance ever made by a lawyer just before his entrance into a trial of a criminal case. His speech comes as a challenge to the defense not to confine the test of the anti-evolution law," Neal asserted, "but instead to put on trial the truth or lack of truth of the theory of evolution [and] the conflict or lack of conflict between science and religion."[109] Such a trial could last up to a month, he predicted. That perfectly fit the defense strategy for the case, and Neal knew that the prosecution opposed it. Yet Bryan's speech offered an opening. When prosecutors did move to exclude such issues from trial, the defense feigned surprise. "We men in New York, when we read the opinion of this distinguished lawyer to the effect that this was a duel to death," Hays protested in court, "we relied then upon the opinion of that distinguished lawyer and we have spent thousands of dollars bringing witnesses here."[110] Of course, those witnesses and both New York lawyers—Hays and Malone—were already en route when Bryan issued his challenge.

"You don't know how glad I am to see you folks!" Rappleyea exclaimed when the two New York lawyers stepped off the train in Chattanooga a day before trial. "Things have been mighty lonesome down at Dayton since Bryan arrived. Mr. Neal, Scopes, and myself have been feeling like three lost kittens." Malone laughed back, "Am I too late for the trial? I rather suspected that I was. You see, I have been reading Mr. Bryan's speeches in the newspapers and I thought the trial had already begun." It immediately became apparent that Bryan no longer had the stage to himself. "The issue is not between science and religion, as some would have us believe. The real issue is between science and Bryanism," Malone added. "I believe that the scientists we have called to act as witnesses in the trial really know more about science than Mr. Bryan; and I also believe that the ministers we have called know more about religion than he does." The prosecution wanted to keep those witnesses off the stand and let Bryan hold forth outside the courtroom. The defense would push its message in and out of court. "The fundamentalists cannot make the issue too broad for us," one reporter eagerly replied to Malone. "The broader they make it, the better we will be satisfied." Hays smiled. "This trial is going to be a good education for the people," he promised, "and for the newspapers."[111]

These comments received wide circulation because a half dozen reporters from major northeastern newspapers rode on the train to Tennessee with Malone and Hays and reported their remarks. Otherwise, in marked contrast to Bryan's reception, the New York defense attorneys arrived without fanfare. "Unknown and unannounced," one reporter noted, "the little group passed quietly through the station to Dr. Rappleyea's car." The only excitement occurred after their arrival in Dayton. Charles Francis Potter, who accompanied Malone and Hays on the trip, became alarmed when a young man grabbed their baggage out of the open trunk. "Hey, boy, what are you doing with those suitcases?" Potter shouted. "That's all right, Doc," Rappleyea replied. "That's only Scopes."[112] The defenders, along with everyone else, had forgotten the defendant.

Darrow arrived later that day, on the last train into Dayton. "There was no torchlight parade to greet me," Darrow later recalled. "Still, there were some people at the depot to meet me; I was received most kindly and courteously at that."[113] Movie cameras captured the scene as Scopes embraced Darrow at the depot—providing the opening footage for newsreels shown throughout the country during the trial. "Scopes is not on trial. Civilization is on trial," Darrow said upon leaving for Dayton. "Nothing will satisfy us but broad victory, a knockout which will have an everlasting precedent to prove that America is founded on liberty and not on narrow, mean, intolerable and brainless prejudice of soulless religio-maniacs."[114] Darrow would stand for individual liberty against mindless majoritarianism—and give no quarter to Bryan. Both sides had worked themselves to a fever pitch. Judge Raulston closed the final day before trial with a remarkable public benediction, as an open-air prayer service occurred on the courthouse lawn. "I am concerned that those connected with this investigation shall divest themselves of all ambition to establish any particular theory for personal gratification," he noted in a public statement. "I am much interested that the unerring hand of Him who is the Author of all truth and justice should direct every official act of mine."[115] Jury selection began the next day.

CHAPTER SIX

PRELIMINARY ROUNDS

THE CROWD gathered early on Friday, July 10, for the opening of the trial. The first spectators began filtering into the courthouse before 7 A.M., a full two hours before the scheduled start. "The newspapermen set along the three sides of the rectangular rail surrounding the sanctum of the court," one of them noted. "Feature writers and magazine contributors have the first three of four spectators' seats reserved for them, just like the seats for the families at a wedding."[1] By 8:45, all seats were taken, and the general public began to spill out into the hallway—local men mostly, from Dayton and the surrounding countryside. "Farmers in overalls from the hillside farms, silent, gaunt men," the *New York Times* reported. "They occupied every seat and stood in the aisles and around the walls of the room."[2] These were not the big-spending tourists that Dayton civic boosters hoped to attract (those people never showed up) but East Tennesseans who came for the day in small automobiles raised high for the rocky mountain roads, or in wagons drawn by horses and mules.

Only about five hundred visitors stayed in Dayton during the trial, and almost half of these were associated with the media. "They sleep and they 'drop' a little money," the *Chattanooga Times* said of the visiting journalists, "but they do not form the vast hoards that Dayton expected."[3] The reporters began work early on the opening day. Some claimed seats at the courtroom press table long before the trial started, already drafting articles for afternoon papers. Antievolution bill author J. W. Butler, in his new role as a trial commentator for a national news syndicate, joined the reporters at the press table—but gave far more interviews than he conducted. Other journalists tracked down Darrow or Bryan for pretrial statements. Press photographers and newsreel crews waited on the courthouse lawn to record the arrival of key participants, much like at a movie premiere. The trial quickly became more of a media event than a spectator show, with some of America's finest journalists on hand to tell the story: H. L. Mencken, Watson Davis, Joseph Wood Krutch, Russell D. Owen, Jack Lait, and Philip Kinsley. Reports from Dayton would dominate the front pages of the nation's newspapers for more than a week.

Judge Raulston arrived with his entire family at about 8:30, carrying a Bible and a statute book. "As he laid these down on the desk," Darrow later wrote, "I wondered why he thought that he would need the statutes. To the end of the trial I did not know."[4] The judge's family took seats next to the bench while the judge mingled with friends and reporters. He wore a new suit for the occasion (Tennessee circuit judges rarely wore robes), and kept the coat on for the time being. With temperatures forecast to push 100 degrees and poor air circulation in the overcrowded courtroom, however, he had authorized attorneys and court personnel to dispense with coats and ties. Most welcomed this relaxed formality, but objected to Raulston's added rule against smoking during the proceedings. Nearly all the lawyers except Bryan smoked heavily. So did the reporters. Perhaps this might be a short trial after all, some people joked. Certainly chewing tobacco gained popularity among court personnel and spectators as a result. A bouquet of flowers graced the judge's bench; spittoons adorned the floor.

Defense counsel came in next, along with Scopes and Rappleyea. Darrow had eaten breakfast with reporters at the Mansion and passed through the gathering crowds on his way downtown, picking up the other defense lawyers as he went. "As we approached the courthouse," Hays later recalled, "our attention was first caught by a sign on the fence reading, 'Sweethearts, come to Jesus,' and conveying other advice of like kind. In the courtyard were various groups of people, some singing psalms."[5] Malone entered the jammed courtroom first, at about

8:45, wearing a fashionable double-breasted suit and smoking a cigarette. "He bubbles with good humor," a New York reporter observed, "and the smile on his round and merry face greets everyone who stops him."[6] Malone was the only lawyer who wore a suit coat throughout the proceedings. At first, townspeople dismissed him as a dandy for doing so, but grew increasingly impressed with his stamina. The rumor spread that he did not even sweat. Although several spectators and prosecutor Ben McKenzie dropped from heat exhaustion during the first day, Malone held out against the temperature in style that day and throughout the trial—although he did occasionally blot his brow with a linen handkerchief.

Darrow trailed behind. "His appearance is in marked contrast to the others of the defense staff," the New York reporter noted. "His huge head, leathery, lined face, square jaw, his twisted mouth of the skeptic, are softened by the quizzical twinkle of his deep-set eyes." Darrow shed his coat upon crossing the threshold, revealing his trademark colored suspenders and pastel shirt—both a generation out of date. "Are you going to wear suspenders like Darrow?" one journalist ribbed Malone, who laughed back, "I refuse to get dressed up for the occasion." Meanwhile Neal impatiently chomped on his half-burnt cigar and Hays chatted with the press. Scopes looked like "a college student on vacation," the reporter added, with neither coat nor tie and his sleeves rolled up to the elbows.[7] His casual dress hid his nervousness. "The whole scene, to me, was unnatural," Scopes later wrote. "I realized that I was on display. Everything I did was likely to be noted; consequently, relaxing was not as easy for me as it apparently was for my companions at the defense table."[8] He made himself as inconspicuous as possible.

From the outset, Judge Raulston adopted the practice—already used by some in town—of referring to Darrow as "colonel." Malone also became "colonel" or the lesser rank of "captain." Darrow submitted to the practice with good humor. Many of the other attorneys also bore titles, but none without an obvious reason. Everyone referred to Neal as "judge," which reflected a previous position; Stewart and Ben McKenzie were usually "general" (as in attorney general); and even Bryan, on occasion, was "colonel," his rank during the Spanish-American War. Yet some wondered whether the judge extended this designation to Darrow and Malone as a way to avoid calling them "mister," a title of respect in the South. He certainly had no problem referring to all the other attorneys, including Hays, as "mister."

The spectators broke into applause just before 9:00, when Bryan entered with Stewart and the other prosecutors. Raulston strode over to welcome the Commoner. The applause broke out anew when Bryan and Darrow shook hands. Despite their differences over religion, the two men had worked together for a variety of political causes and remained on cordial terms. In a letter to Sue Hicks shortly before trial, Bryan described Darrow as "an able man, and, I think, an honest man."[9] Darrow, for his part, always maintained that Bryan was sincere. The two talked amiably with their hands clasped on each other's shoulders and posed for pictures with the judge. The conversation became more formal when Malone approached Bryan, perhaps due to bitterness from their days together at the State Department. Darrow wandered off to compare suspenders with Ben McKenzie.

Bryan already showed signs of strain from the heat. "His shirt sleeves were rolled up as high as they would go, and his soft collar and shirt front were turned back away from his neck," Darrow recalled, though closer inspection showed that Bryan had removed his collar altogether. "In his hand was the largest palmleaf fan that could be found, apparently, with which he fought off the heat waves—and flies."[10] Mary Baird Bryan watched from a wheelchair behind her husband. She suffered through the entire trial in quiet dignity and obvious physical pain from her crippling arthritis. Privately she objected to her husband's crusade against teaching evolution and his participation in the Scopes trial, but she stood by him throughout. Raulston took his seat behind the bench and called for order. The trial was about to begin.

First, the court opened with a long prayer (Scopes called it "interminable") by a local fundamentalist minister. "Not just an ordinary prayer," Hays noted, "but an argumentative one, directed straight at the defense." Acknowledging a divine "source of our wisdom," the preacher prayed that "the Holy Spirit may be with the jury and with the accused and with all the attorneys" so that they would "be loyal to God." Many spectators punctuated these words with audible amens. Prosecutors bowed their heads throughout; reporters looked toward the defense table; the defense lawyers stared out the window. The judge then reconvened the same grand jury that had indicted Scopes six weeks earlier; that jury had met in May without sufficient notice, therefore a new indictment was needed. The judge repeated his original charge to these jurors, complete with the Genesis account of creation, and Stewart recalled the earlier witnesses. "One of the [student witnesses] did not want to go on the stand," Scopes later wrote. "To prevent his loyalty [to me] from delaying the trial I went to see the youngster and told him to go ahead . . . because he would be doing me a favor." These proceedings consumed most of the morning, whereupon counsel asked to call it a day so that defense lawyers could recuperate from their travels and become acclimated to the heat. "Well, it wouldn't require any great amount of energy to select a jury, would it?" the judge responded. He then directed the sheriff to summon one hundred potential jurors to appear after lunch. Under local practice, all veniremen would be white males.[11]

Shortly before noon, a thousand people poured out of the ovenlike

The principal figures in the Scopes trial greeting each other in the crowded courtroom at the start of the trial. In front from left are Dudley Field Malone, Tom Stewart, William Jennings Bryan, Judge John Raulston, and Clarence Darrow. (Courtesy of Bryan College Archives)

courtroom into the festive atmosphere of downtown Dayton. Four steers roasted over a huge barbecue pit behind the courthouse. Hot dog and soft drink stands lined the main street, intermixed with bookstalls and carnival games. "A blind man with a portable organ sat at the iron fence at Market Street, only half shaded from the broiling sun, playing mountain hymns," a reporter observed, "another blind man played on a guitar and mouth organ." An African-American string quartet entertained in the street. "Negroes mingled freely with white persons on the lawn of the court house," a surprised Yankee noted. The biggest thrill occurred later in the day, when two airplanes buzzed the crowd after taking off from near town. They carried newsreel footage of the trial that would begin showing in northern cinemas at the next afternoon's matinee.

Jury selection started immediately after lunch. Darrow typically stressed this part of a trial as critical for the defense and often spent weeks going through hundreds of veniremen before settling on twelve suitable jurors. Tennessee trial practice allowed only three peremptory challenges without cause for each side, and there was little point in probing into the backgrounds and beliefs of veniremen to establish cause for their exclusion owing to a fundamentalist predisposition—which in itself would never constitute just cause for a local judge to exclude anyone from anything. When Darrow tried this tact by challenging a particularly militant fundamentalist for cause, Stewart objected: "If a man is subject to challenge by the defendant because he believes the Bible conflicts with the theory of evolution . . . then, for the converse reason the state would have grounds to challenge for cause and the result would be everybody on earth who could be brought here, would be challenged." Betraying his frustration with the jury pool, Darrow shot back, "If you can find any man on the jury that believes in evolution, you have my permission to challenge him."[12] Despite Stewart's objection, this particular venireman had gone too far in admitting his bias. Raulston excused him for cause.

Darrow settled for jurors who claimed to have an open mind. To facilitate this, he asked that names from the jury pool be drawn from a hat rather than selected by the sheriff. The judge offered this option to accommodate Darrow's concern about fairness, inviting his daughter to draw the first name. The prosecution readily accepted nearly everyone after a few pro forma questions. For the defense, Darrow engaged each potential juror in a casual interrogation that inevitably covered three key issues and generally elicited similar responses. "Mr. Smith, do you know anything about evolution?" Darrow began one typical exchange. "I do not, no sir," came the inevitable reply. Further questioning led up to, "Did you ever have any opinion . . . on whether the Bible was against evolution or not?" No, again. Finally, Darrow inquired to the effect, would you make up your own mind on these matters based on the evidence presented in court? When the answer came back, "Yes, sir," Darrow concluded, "I think you would, too. You are a juror."[13]

Nearly every venireman wanted to join the jury, if for no other reason than that it appeared to offer a front-row seat for the proceedings. Clearly, many of them said whatever they thought would help to get them accepted. This typically included denying that they held any opinion regarding the theory of evolution and its relationship to Christianity. Some were transparently honest in their professed ignorance on these points. When asked if he had ever read about the subject, for example, one venireman replied, "I can't read." Darrow followed up, "Is that due to your eyes?" No, the man answered, "I am uneducated." Hays later commented, "It was said with such plain, simple dignity that we felt we had at least one honest man." The illiterate venireman joined the jury. "Evolution is a new idea to the average Tennessee juryman," Watson Davis concluded. This gave the defense grounds for arguing that the jurors needed to hear expert testimony about evolution to decide the case.[14]

When pushed, however, these veniremen betrayed a marked fundamentalist tilt. None said anything negative about the Bible or positive about evolution, and all but one of them were church members. Most were middle-aged farmers from rural Rhea County with little formal education. The judge excluded a few for cause after Darrow probed deeper into their beliefs. For example, a rural minister who professed to know nothing about evolution aroused Darrow's suspicions. In response to rapid-fire grilling, the minister first denied ever preaching about evolution, then admitted doing so "in connection with other subjects," and finally exclaimed, "Well, I preached against it, of course!"[15] Local spectators broke into loud cheers, but the minister lost his chance to sit in judgment of Scopes.

Usually Darrow took prospective jurors at their word and accepted the inevitable. "It was obvious after a few rounds that the jury would be unanimously hot for Genesis. The most that Mr. Darrow could hope for was to sneak in a few men bold enough to declare publicly that they would have to hear the evidence against Scopes before condemning him," H. L. Mencken reported from the scene. "Such a jury, in a legal sense, may be fair," he added, "but it would certainly be spitting in the eye of reason to call it impartial."[16] The entire process took only two hours and twenty veniremen. Darrow told reporters afterward, "It is as we expected." Bryan commented for the prosecution, "We are satisfied."[17] Many northern editorialists scorned the prospect of these jurors sitting in judgment on a scientific theory, but one put it in a larger perspective. "Last week the white press made much ado about the jury that now sits hearing evidence in the Scopes' trial," a Pittsburgh African-American newspaper editor noted. "But right here we rise to remind the complainants that this is no unusual spectacle. The Scopes' jury is typical—typical of the judgment bar before which black men and women in the bourbon South must stand when charged with crime against members of the opposite group."[18] Scopes now stood charged with such a crime, and Bryan's majority sat in judgment.

Jury selection concluded quicker than anyone expected, and the court prepared to adjourn early for the defense. Counsel raised one more issue: "That is the matter with reference to the competence of evidence that will be introduced by the bringing here of these scientists," Stewart spit out.[19] Darrow had raised the issue several times during the day. Indeed, in his first words to the court that morning, Darrow said, "Your honor, before [jury selection] I want to have a little talk . . . on the question of witnesses here, before we do anything else." Raulston put him off. The competency of defense witnesses usually would not come up until the prosecution concluded its case, and the defense offered those witnesses, but Darrow pushed for an early ruling. "Your honor, all I am doing at this time is because our witnesses are generally from a long distance," he stated. "If there is to be any question of competency of evidence, that could be disposed of some time before we get them here." The prosecution left no doubt about its position. "We have had a conference or two about that matter," Stewart replied. "It isn't competent to bring into this case scientists who testify as to what the theory of evolution is or interpret the Bible or anything of that sort."[20] Yet things of that sort constituted the *entire* defense.

"Education is the real job of the defense," Watson Davis reported from Dayton that first day. He assumed responsibility for assembling defense witnesses, assuring readers that the "supply of competent and learned professors will be ample. Dayton may be the scene of upward steps in the evolution of the human thinking mind. Perhaps that is not too much to hope for."[21] Of course, it was too much to count on. Governor Peay and other prominent Tennesseans already had warned the court against admitting expert testimony and prosecutors vowed to oppose it as irrelevant. Stewart suggested that the prosecution would agree to take up the competency issue next. Judge Raulston offered to hear the matter on Saturday, so that the defense would "have the advantage over Sunday to arrange for witnesses or not," but Scopes's travel-weary defenders asked to wait until Monday.[22] The prosecution did not object and court adjourned for the weekend. Bryan had not spoken in court on this, the twenty-ninth anniversary of his famous "Cross of Gold" speech; he left legal matters to the other prosecutors, and did not intend to address the court until closing arguments—when he planned to expose the menace of Darwinism to the country in a carefully crafted oration.

"Dayton is having a case of the morning after today," Jack Lait of William Randolph Hearst's International News Service (INS) reported on Saturday. "In numbers, [the opening day] was a fiasco; the procedure lacked drama; and then came the forty-eight-hour adjournment to let what warmth that had radiated cool off."[23] Dayton quieted down for the weekend. Trial spectators from the surrounding countryside returned to their homes. Most visitors from outside Tennessee headed to Chattanooga for a hot time or the Great Smoky Mountains for a cool breeze. Reporters and defense lawyers enjoyed a riverboat excursion on the Tennessee River, compliments of the *Chattanooga News*. Prosecutors drove into the mountains for the day. About the only excitement occurred when a self-proclaimed Independent Free Thinker and Lecturer began loudly assailing Christianity on a downtown street corner. He was arrested for disturbing the peace and released on condition that he stop speaking in public.[24] To prevent similar outbursts, town officials denied permission for another itinerant agnostic to speak from the platform on the courthouse lawn and closed the area to all speakers two days later. "It would be hard to image a more moral town than Dayton. If it has any bootleggers, no visitor has heard of them," Mencken wrote after his first weekend in the community. "No fancy woman has been seen in the town since the end of the McKinley administration. There is no gambling. There is no place to dance."[25] The Saturday night jazz party occurred six miles outside Dayton, at the Morgan Springs Hotel.

Bryan did not join the other prosecutors on their excursion, but stayed behind working for the cause. One newspaper referred to him as the prosecution's "loud speaker"—he broadly attacked teaching evolution outside the courtroom while Stewart narrowly defended the antievolution statute inside it.[26] Bryan began the day by issuing a statement endorsing Stewart's decision to oppose the introduction of expert testimony. "If the people of Tennessee have a right to pass laws for the protection of the religion of their children, then they have a right to determine for themselves what they consider injurious," the Commoner reasoned. "No specialists from the outside are required to inform the parents of Tennessee as to what is harmful."[27] He spent most of the afternoon sitting in the shade of a maple tree in his front yard greeting well-wishers and preparing two speeches for the next day. Just in case the court admitted expert witnesses for the defense, Bryan wired Straton, Riley, and Norris to stand ready to appear on short notice if their testimony was needed for rebuttal by the prosecution.

Reporters promptly carried Bryan's early morning statement to Darrow, who still resided at the Mansion. He responded to reporters from his bed, which was about the only place in his room to sit. Darrow rehearsed the defense's contention that, owing to the statute's wording, the prosecution must prove not only that Scopes taught about human evolution but also that such teaching denied the biblical account of creation. Darrow, of course, believed that the theory of evolution flatly contradicted the Bible, but the defense planned to present Christian scientists and theologians who professed otherwise. "Mr. Bryan says [jurors] would decide all this without evidence. It is obvious that no jury can accomplish such a thing," Darrow declared. "Whether the scientists come from Tennessee or outside to tell the meaning of evolution can not matter. Science is the same everywhere. The Constitution does not permit the legislature to put a Chinese wall around the state."[28]

Darrow became even angrier by late afternoon when he heard that the prosecution no longer would agree to an expedited hearing on the issue of expert testimony. Stewart had come back from the mountains worried that deciding the issue out of order might constitute procedural error as an advisory opinion by the court. He took his concern to defense attorneys at the Mansion, where Hays and Neal concurred with him; Darrow heard about it after Stewart left and strenuously objected, but it was too late. The Chicago attorney was "fighting mad," the *New York Times* reported: "We will try to hold [prosecutors] to their agreement, but of course cannot do so if they persist," the defense announced.[29]

Tensions mounted further on Sunday, when Bryan took to the stump. He began the day by delivering the morning sermon to a packed house at Dayton's southern Methodist church. Bryan now answered Darrow's latest statement. "The attorneys for the defense charge that our objection to expert testimony is an attempt to evade the issue. On the contrary it is an effort to confine the case to the issue," he asserted. "The statute itself distinctly forbids the teaching of the evolutionary hypothesis"—regardless of whether or not it conflicts with the Bible. "Then, too, their testimony would necessarily be one-sided," he added in a comment that spoke to the nature of America's adversarial judicial system. "They will only call those who still cling to religion and try to harmonize evolution with it. They will thus present a very one-sided view of evolution and its results. A half truth is sometimes worse than a lie, and evolution as they want to present it is less than a half truth."[30] Judge Raulston and his entire family sat in the front pew as the congregation cheered the Commoner. At the same hour, Charles Francis Potter's plan to address Dayton's northern Methodist church on the topic of evolution and religion fell through due to objections from the congregation. Adding to local complaints against evolutionists, the morning paper reported a proposal to bar Tennessee public school graduates from Columbia University, prompting Superintendent White to suggest that Dayton found its own university—and to name it for Bryan.

In midafternoon, Bryan delivered a prepared speech from the speaker's platform on the courthouse lawn to a crowd of some 3,000 people—most of whom came from Dayton and the surrounding countryside. Town regulations now forbad platform speakers from discussing evolution, but Bryan got his point across. "When the schools get through with our children they must still have something else," he proclaimed: Values, which come from faith in God. "Mr. Bryan's manner with these people is most persuasive. His voice seems to reach out and caress them," the *New York Times* reported. On the periphery of the crowd, a member of the defense team complained to a journalist,

"What is that but an attempt to influence the judgment of the community?" Potter planned to answer Bryan from the same platform that evening, but T. T. Martin held a permit to preach at that site every evening during the trial, with the sole restriction that he not discuss evolution. After some discussion, Martin deferred to Potter this one time on the condition that the Unitarian obey the same restriction. Potter complied by delivering an uninspiring plea for liberal education as the basis for values. The defense, however, chomped at the bit to regain the offense once court resumed on Monday.

Once again the out-of-town press and local spectators jammed the cavernous courtroom. According to an observer, "The crowd filled the aisles, the windows, the doors, the space behind bar and bench, while photographers and movie men were perched on chairs, tables and ladders." Townspeople appeared to replace country folk in the gallery, and women now attended in approximately equal numbers as men.[31] The judge delayed the call to order for a quarter hour as press photographers snapped pictures and the radio announcer tested microphones. Darrow informally asked the judge to dispense with courtroom prayers, "particularly as the case had a religious aspect."[32] Raulston dismissed the request and invited forward another conservative local pastor, who jabbed the defense with a prayer to God as "the creator of the heaven and the earth and the sea and all that is in them."[33] Over the weekend, the county had installed three portable electric fans to circulate the air, but they provided little relief and the cords did not reach anywhere near the defense table. Courtroom theatrics helped to distract spectators, however, as the tension mounted with the temperature.

Even without consideration of the competency of expert testimony, this was a crucial day for the defense. It presented an opportunity, at the start of trial, to challenge the constitutionality of the antievolution statute through a motion to quash the indictment. To preserve all conceivable issues for appeal, the formal motion to quash identified fourteen separate constitutional objections to the statute, but many of these contained so little substance that the defense never mentioned them in their oral arguments. Most of the serious objections invoked provisions of the Tennessee state constitution because the federal Bill of Rights then had limited application against states. The key state constitutional provisions included express guarantees of individual freedom of speech and religion, requirements for clearly understandable indictments and titles for legislation, and a clause directing the legislature to cherish science and education. Furthermore, both the Tennessee and United States constitutions barred the state from depriving any person of liberty without due process of law, which courts then interpreted as precluding patently unreasonable state laws and actions.

The defense took this motion seriously. At the outset, Neal asked the court to confirm the procedure for debating the motion: "We would have the right to make an explanative statement and then the Attorney-General makes his argument, and we to make the final argument." If the court later excluded expert testimony, this might offer the defense's best opportunity to present its case in court. Neal and Hays would lay the foundation, but they wanted to save the dramatic closing for Darrow—leaving no chance for Bryan to rebut it. The judge agreed.[34] The jury was excluded throughout because, under established practice, the judge decided legal questions regarding the constitutionality of the statute and the validity of the indictment. If both passed muster, the jury then determined the defendant's guilt under them. So much for the jurors' front-row seats. They left the courtroom early Monday and did not return until Wednesday afternoon.

Neal opened by reading a rambling commentary touching on the major constitutional issues raised by the motion. Unwashed and unshaved as usual, he lectured the court in a manner reminiscent of his chaotic classroom teaching style. During his presentation, Neal returned most often to the constitutional bar against the establishment of religion in public schools, asserting that "the legislature spoke for the majority of the people of Tennessee [in passing the antievolution law], but we represent the minority, the minority that is protected by this great provision in our constitution."[35]

Hays followed with a more coherent statement that focused on one issue: the reasonableness of the statute as an exercise of police power under the due process clauses of the state and federal constitutions. "My contention is that no law can be constitutional unless it is within the right of the state under the police power, and it would only be within the right of the state to pass it if it were reasonable," he maintained. To illustrate the statute's unreasonableness, Hays compared it to a hypothetical law against teaching that the earth revolved around the sun. "My contention is that an act of that sort is clearly unconstitutional," he explained, "and the only reason Your Honor would draw a distinction between the proposed act and the one before us is that . . . the Copernican theory is so well established that it is a matter of common knowledge." But, Hays asserted, "Evolution is as much a scientific fact as the Copernican theory." Scientific expertise rather than common knowledge should set the standard of reasonableness for science education. "Of course," Hays later explained in recapping his argument, "the State may determine what subjects shall be taught, but if biology is to be taught, it cannot be demanded that it be taught falsely."[36]

The prosecution countered with arguments by the past and present attorneys general for Dayton's judicial district, Ben McKenzie and Tom Stewart, two prosecutors with sharply contrasting styles. McKenzie began practicing law before Stewart's birth; he personified the stereotypical old-style southern politician—a windbag blowing forth flowery speech, meaningless compliments, and folksy humor. McKenzie began by rejecting Hays's analogy to a law against Copernicanism: "It is not half so much kin to this case as he says we are to the monkeys." When the laughter died down, he added in his high, raspy voice, "No such act ever passed through the fertile brain of a Tennessean." More laughter led him to add, "But I don't know what might happen up in his country." Looking out over reading glasses with a twinkle in his eye, the elderly Dayton attorney followed a similar approach in defending the statute's clarity. "The smallest boy in our Rhea County Schools could understand it," he joked. "We don't need anybody from New York to come down here to tell us what it means." Malone broke the spell by objecting to the "geographical" slurs. "Why you all ain't acquainted with me. I love you," McKenzie replied in a broad southern drawl. "And I love you, but I want you to stick to the motion," Malone snapped back in his clipped New York Irish accent. "I love you," McKenzie insisted. "Sure you do," Darrow added.[37]

Stewart took over for the prosecution after lunch. He represented a new generation of southern politician—ready, willing, and able to compete on an equal basis with the best northern attorneys. He was not a fundamentalist and questioned the wisdom of the antievolution law, but he took pride in southern culture, including its Protestant religious traditions, and wholeheartedly defended the legislature's constitutional authority to adopt the challenged statute. When defense counsel interrupted his argument by asking if the statute favored Christianity over other religions, Stewart matter-of-factly dismissed their question as absurd. "The laws of the land recognize the Bible," he answered. "We are not living in a heathen country."[38]

Stewart kept returning to his main point regarding the statute: "It is an effort on the part of the legislature to control the expenditure of state funds, which it has the right to do." Individual freedom was not at stake.

Chief prosecutor Tom Stewart during a break in the Scopes trial. (Courtesy of Bryan College Archives)

"Mr. Scopes might have taken his stand on the street corners and expounded until he became hoarse," Stewart asserted, "but he cannot go into the public schools . . . and teach his theory." Legislators, "who are responsible to their constituents, to the citizens of Tennessee," should control public education. Like Bryan, Stewart stressed majority rule; unlike Bryan, he never bashed evolution. Lawmakers could exclude *any* subject from the public school curriculum according to Stewart, and he cited ample legal precedents to support this general assertion.[39] "Many leaving the courtroom were heard to say that the 33-year-old attorney-general, in his clashes with the veterans of the opposing counsel, 'took pretty good care of himself,'" one reporter observed.[40]

By that time, however, most spectators were discussing Darrow's brilliant rebuttal. "I made a complete and aggressive opening of the case," Darrow later explained. "I did this for the reason that we never at any stage intended to make any [closing] arguments in the case." In that upcoming case to the jury, which would follow debate on the pending motion, the prosecution would offer its evidence that Scopes taught about human evolution and the defense would try to introduce expert testimony on science and religion. At that point, the defense planned to waive closing arguments and submit the matter to the jury. Bryan spent weeks preparing his closing arguments for the trial. It would have come at the end of the case, with no chance for a courtroom response. Defense attorneys feared its impact on the jury and the public. "By not making a closing argument on our side we could cut him out," Darrow explained.[41] Such a trial-ending tactic placed tremendous importance on his beginning plea to quash the indictment. Darrow might not get another chance to state his case in court. Furthermore, by saving his argument on the motion to quash for the rebuttal, and thereby becoming the last speaker on that opening issue, the prosecution could not respond to it. Darrow rose to the occasion.

"Clarence Darrow," the *New York Times* proclaimed in its lead story, "bearded the lion of Fundamentalism today, faced William Jennings Bryan and a court room filled with believers of the literal word of the Bible and with a hunch of shoulders and a thumb in his suspenders defied every belief they hold sacred."[42] As with many powerful speeches, the argument was simple yet delivered with great impact. "We have been informed that the legislature has the right to prescribe the course of study in the public schools. Within reason, they no doubt have, no doubt," Darrow began. But "the people of Tennessee adopted a constitution, and they made it broad and plain, and said that the people of Tennessee should always enjoy religious freedom in its broadest terms, so I assume that no legislature could fix a course of study which violated that."[43] He had answered Stewart.

Darrow's opening introduced his main point. The antievolution statute was illegal because it established a particular religious viewpoint in the public schools. Darrow presented this defense in state constitutional terms because the U.S. Supreme Court had not yet interpreted the Constitution's establishment clause to limit state laws—but otherwise both state and federal constitutions offered similar protections. He began reading from and commenting on the Tennessee constitution: "'All men have a natural and indefeasible right to worship Almighty God according to the dictates of their own conscience.' That takes care even of the despised modernist, who dares to be intelligent." He resumed reading, "and that 'no preference shall be given by law to any religious establishment or mode of worship.' Does it? Could you get any more preference, your honor, by law?" Darrow explained, "Here is the state of Tennessee going along its own business, teaching evolution for years." He turned toward Bryan. "And along comes somebody who says we have to believe it as I believe it. It is a crime to know more than I know. And they publish a law inhibiting learning." That law established a religious standard, Darrow charged: "It makes the Bible the yard stick to measure every man's intellect, to measure every man's intelligence and to measure every man's learning." Bryan "is responsible for this foolish, mischievous and wicked act," Darrow thundered. "Nothing was heard of all that until the fundamentalists got into Tennessee."[44]

Darrow spoke in dead earnest, expressing a liberal skeptic's view of religion. Hundreds of creeds existed within Christianity alone, he noted, not to mention all the other religions of the world. "The state of Tennessee under an honest and fair interpretation of the constitution has no more right to teach the Bible as the divine book than that the Koran is one, or the book of Mormon, or the book of Confucius, or the Buddha, or the Essays of Emerson," he snarled. "There is nothing else, your Honor, that has caused the difference of opinion, of bitterness, or hatred, of war, of cruelty, that religion has caused." Darrow quoted the maxim, "To strangle puppies is good when they grow up into mad dogs," and suggested that it applied to fundamentalism, which threatened "to kindle religious bigotry and hate" in America. The Bible itself contained differing accounts of creation, Darrow added. "It is not a book on biology, [its writers] knew nothing about it. . . . They thought the earth was created 4,004 years before the Christian Era. We know better. I doubt if there is a person in Tennessee who does not know better." Most intelligent Christians accepted the theory of evolution too, he asserted, "and that the God in which they believe did not finish creation on the first day, but that he is still working to make something better and higher still out of human beings." Bigotry, ignorance, and hatred marked the antievolution crusade according to Darrow, "But your life and my life and the life of every American citizen depends after all upon the tolerance and forbearance of his fellowman."[45]

"While he was talking there was absolute silence in the room except for the clicking of telegraph keys," the *New York Times* reported. "His words fell with crushing force, his satire dropped with sledgehammer effect upon those who heard him."[46] H. L. Mencken added, "You have but a dim notion of it who have only read it. It was not designed for reading but for hearing. The clangorousness of it was as important as the logic. It rose like a wind and ended like a flourish of bugles."[47] Darrow paced as he spoke, tugging on his lavender suspenders. "He would stop and brood a minute, hunching his shoulders almost up to his ears, and then they would drop, his head would shoot forward and his lower lip protrude as he hurled some bitter word at his opponents," the *New York Times* noted.[48] The *Chicago Tribune*, Darrow's hometown paper, classed it as "one of the greatest speeches of his career."[49] It continued for two hours until the judge interrupted at the prescribed time for adjournment. Even then Darrow persisted for another ten minutes. "There was unquestioned greatness both in the passion with which it was uttered and in the calculation of the moment for utterance," Joseph Wood Krutch wrote in *The Nation*, "and when [Darrow] concluded with the solemn warning that 'we are marching backwards to the glorious age of the sixteenth century when bigots lighted fagots to burn men who dared to bring any intelligence and enlightenment and culture to the human mind,' even Dayton stopped to think."[50]

Telegraphs transmitted 200,000 words from Dayton that day, a record for a single event. Newspapers across the country reprinted Darrow's speech at length, and many editors echoed his plea for tolerance.

Defense attorneys rushed forward to congratulate Darrow. "We looked upon the day's work and found it good," Hays later commented, "a ray of light had been flashed in Tennessee."[51] Ben McKenzie extended effusive compliments, hailing Darrow's effort as "the greatest speech that I have ever heard on any subject in my life." Ruby Darrow proudly fussed over her husband's sleeve, which had split open during one gesture.

Not everyone in the courtroom had the same reaction. Some spectators hissed at the end (Mencken called them "morons"), and one remarked, "They ought to put him out!"[52] Bryan—coatless, collarless, and wet with perspiration—sat silently throughout, trying to cool himself and shoo flies with his palmleaf fan. "Somehow," Darrow later recalled, "he did not look like a hero. Or even a Commoner. He looked like a commonplace fly-catcher."[53] The Memphis *Commercial Appeal* captured much of the local sentiment in a front-page editorial cartoon picturing a cold, aloof Darrow huddled atop a black mountain in hell, surrounded by skulls of "annihilation," the dragon of "agnosticism," and a bowed prisoner of Satan labeled "spiritual despair." The caption read, "Darrow's Paradise!"[54] A day later Mencken reported, "The net effect of Clarence Darrow's great speech yesterday seems to be precisely the same as if he had bawled it up a rainspout in the interior of Afghanistan. That is, locally, upon the process against the infidel Scopes, upon the so-called minds of these fundamentalists of upland Tennessee."[55] Hays agreed: "Personally, I doubt whether at any time the attorneys for the prosecution caught our point on the religion question. Every word, to say nothing of emotions, in court made it clear that there was really no other question."[56]

The court reconvened on time the next morning, but promptly adjourned until afternoon. A powerful storm, which some visiting reporters jokingly attributed to divine displeasure with Darrow's speech, had disrupted the town's power and water on Monday night. As a result, Raulston had not finished preparing his ruling on the motion to quash the indictment. He needed a few more hours. In the meantime, the only official business before the court consisted of the opening prayer. To highlight its contention that the case raised a religious question, and thereby to underscore its establishment clause argument, the defense now formally objected to public prayer in the courtroom. "When it is claimed by the state that there is a conflict between science and religion," Darrow stated, "there should be no . . . attempt by means of prayer . . . to influence the deliberations." Ben McKenzie defended the practice by citing a state supreme court decision that permitted voluntary prayer by jurors. Darrow responded by drawing a modern-day distinction between public and private religion: "I do not object to the jury or anyone else praying in secret or private, but I do object to the turning of this courtroom into a meeting house."[57]

Editorial cartoon during the Scopes trial presenting popular view of Darrow's militant agnosticism. (Copyright © 1925 the *Commercial Appeal,* Memphis, Tennessee. Used with permission)

Stewart had heard enough. He did not want to lose control of the proceedings. Sensing Darrow's strategic objective in raising the objection, Stewart promptly denied that any religious question existed in the case. "It is a case involving the fact as to whether or not a schoolteacher has taught a doctrine prohibited by statute," he asserted, again avoiding any mention of evolution. Stewart also rejected Darrow's views on the inappropriateness of public prayer, stating that "such an idea extended by the agnostic counsel for the defense is foreign to the thoughts and ideas of the people who do not know anything about infidelity and care less." Darrow fixed his deep-set eyes directly on the fiery young prosecutor who, according to one reporter, "was trembling with suppressed emotion as he forced out his last words." The judge tried to defuse the situation, pleading at one point, "Gentlemen, do not turn this into an argument," but overruled the objection. The prayer was heard.[58]

The issue of prayer resurfaced when court resumed in the afternoon. The defense again drew attention to the religious issue underlying the case by submitting a petition to the court, signed by Potter and other visiting modernist clerics, asking that "clergymen from other than fundamentalist churches" alternately deliver the opening prayer. Hays then moved that "we have an opportunity to hear prayers by men who think that God has shown His divinity in the wonders of nature, in the book of nature, quite as much as in the book of the revealed word." Perhaps no single sentence during the entire trial better captured the difference between modernism and fundamentalism.[59] Judge Raulston deftly referred the petition to the local pastors' association and asked that group to choose who should deliver future courtroom prayers. Visiting journalists began to laugh. Local spectators cheered. Hays objected. Everyone thought that this would preclude modernists from the task, but the association picked Potter for the very next day and alternated between fundamentalists and modernists thereafter.

Tensions reached a high point that afternoon. Spectators filled every available seat, and several hundred people stood in the aisles and along the walls. County officials worried aloud that the floor might collapse under the weight. Power and water remained out, stopping the electric fans and drinking water. Nothing happened. "For a hour and a half," the *Commercial Appeal* reported, "the hot, bustling crowd puffed, fanned, smoked and drank red soda pop waiting for the judge."[60]

Scopes lit one cigarette after another. A rumor spread that an INS reporter had scooped the judge's ruling. Raulston finally entered the courtroom at 3:45 and after dealing with the clerics' petition, addressed the press. "I am now informed that the newspapers in the large cities are now being sold which undertake to state what my opinion is," he sternly lectured. "Now any person that sent out any such information as that, sent it out without the authority of this court and if I find that they have corruptly secured such information I shall deal with them as the law directs." The judge adjourned court for the day without issuing his ruling and appointed a committee of five leading journalists to investigate the source of the leak. "Judge Raulston was very angry," one of the reporters noted, "and ready to take severe measures with any culprit; the newspaper men were split in rival camps and at dagger's points with one another."[61]

Tempers cooled slowly. Stewart stomped out of the courtroom still angry over the clerics' petition. "What does he think this is, a political convention?" the chief prosecutor asked reporters. "It's going to be mighty rough from now on," Malone warned. In a press interview, Bryan grimly concluded that "this case uncovered a concerted attack upon revealed religion that is being made by a minority made up of atheists, agnostics, and unbelievers."[62] Some of the assembled journalists could not help joking about the scooped story. Fortified with bootleg liquor, the press committee that evening conducted a mock trial of the young INS correspondent who had written the offending article. Amid howls of laughter, the young reporter identified his source as the judge. Apparently Raulston inadvertently tipped his hand on his way to lunch when, in response to the correspondent's questions, he agreed that the trial would resume after the ruling, thereby implying that the indictment would stand.

The lighter mood carried over into the courtroom on Wednesday and the trial got back on track. After hearing the report from the press committee, Raulston let off the INS correspondent with a stern lecture then read the long-awaited ruling on the motion to quash the indictment. One by one, the judge rejected defense objections to the statute. On the constitutional issue of religious freedom, he opined, "I cannot conceive how the teachers' rights under this provision of the constitution would be violated by the act in issue. . . . The relations between the teacher and his employer are purely contractual and if his conscience constrains him to teach the evolution theory, he can find opportunities elsewhere."[63] The court had adopted the prosecution's position, which accorded with the prevailing currents of constitutional interpretation. "I don't think anyone was surprised," Scopes later wrote. "No one of the defense had expected Judge Raulston to rule the Butler Act unconstitutional or otherwise to view favorably the motion to quash."[64] The trial proper would begin after lunch.

—CHAPTER SEVEN—

THE TRIAL OF THE CENTURY

WEDNESDAY WAS the hottest day of the trial, or so it seemed to many inside the courthouse. One observer called it "the worst day of all," and complained of "the crowd filling the court rooms so that a breath of air through the windows was almost impossible."[1] Only the renewed cordiality among participants made it tolerable. When prosecutor Ben McKenzie appeared on the verge of collapsing from the heat again, Malone rushed over to fan him. During the noon recess, two young prosecutors, Wallace Haggard and William Bryan, Jr., went swimming with the defendant in a mountain pond. "The water was cool and clear," Scopes later recalled. "We temporarily forgot the trial and everything; as a result we were late returning to the courtroom." When they finally showed up, Scopes could barely squeeze through the packed aisles to the defense table. "Where the hell have you been?" thundered Hays, but no one else appeared to notice the defendant's absence.[2]

Prosecutors had too much trouble locating their own witnesses—schoolboys lost in a sea of adults—to worry about Scopes, and by the time they found them they had lost their chairs to spectators. Ben

McKenzie called on the unknown culprits to return the chairs: "We are a necessary evil in the courtroom," he protested.

When the prosecutors and their witnesses were finally in place, the court recalled the jurors and directed each side to make its opening statement. Stewart earlier predicted that his case would take "about an hour," and kept to that pace by delivering a two-sentence opening statement.[3] Scopes violated the antievolution law by teaching that "mankind is descended from a lower order of animals," the prosecutor simply declared. "Therefore, he has taught a theory which denies the story of divine creation of man as taught in the Bible."[4]

Defense counsel estimated that their experts, if permitted to testify, would talk for weeks, and accordingly countered with an extended, carefully crafted opening statement. "We will prove that whether this statute be constitutional or unconstitutional the Defendant Scopes did not and could not violate it," Malone read from a typed script. "We will show by the testimony of men learned in science and theology that there are millions of people who believe in evolution and in the story of creation as set forth in the Bible and who find no conflict between the two. The defense maintains that this is a matter of faith and interpretation, which each individual must determine for himself." "[S]o that there shall be no mistake," he noted, "the defense believes that there is a direct conflict between the theory of evolution and the theories of creation as set forth in the Book of Genesis," but this simply represented the *opinion* of counsel. "While the defense thinks there is a conflict between evolution and the Old Testament, we believe there is no conflict between evolution and Christianity." Among the defense lawyers, only Malone could have read this line with conviction; it did reflect the beliefs of their modernist Christian expert witnesses, however. Malone suggested succinctly three different views on the relationship between the Bible and evolution: complete accord, direct conflict, and progressive compatibility. Accepting any one viewpoint constituted a matter of personal religious opinion, he asserted. The prosecution simply could not assume, nor could it prove, that teaching the theory of human evolution denied the biblical account of creation.[5]

Malone directly assailed Bryan. "There might be a conflict between evolution and the peculiar ideas of Christianity which are held by Mr. Bryan as the evangelical leader of the prosecution, but we deny that the evangelical leader of the prosecution is an authorized spokesman for the

Christians of the United States," Malone explained. To emphasize the transitory nature of religious opinion, he quoted from a twenty-year-old article in which Bryan endorsed Thomas Jefferson's Statute of Religious Freedom in language that seemingly repudiated laws—such as the antievolution statute—that coerced or promoted religious belief. "The defense appeals from the fundamentalist Bryan of today to the modernist Bryan of yesterday," Malone declared. The repeated references to Bryan finally drew an objection from Stewart, but the Commoner waved him off. "I ask no protection from the court," Bryan asserted, "and when the proper time comes I shall be able to show the gentlemen that I stand today just as I did, but that this has nothing to do with the case at bar."[6] After waiting days for their Peerless Leader to speak in court, the local spectators erupted. "They stamped. They whistled, they cheered with their lungs. They applauded with their hands," one reporter observed. "Bryan had won. His simple eloquence had confounded the sophistry of his enemies."[7] Watching this demonstration, Mencken observed, "Bryan is no longer thought of as a politician and jobseeker in these Godly regions, but has become converted into a great sacerdotal figure, half man and half archangel—in brief, a sort of fundamentalist pope."[8]

After order was restored and Malone finished reading his opening statement, the prosecution expeditiously presented its case. Stewart called four witnesses. Superintendent White led off by testifying that Scopes had admitted teaching about the theory of human evolution from Hunter's *Civic Biology* when conducting a review session for a high school biology class. Stewart identified offending paragraphs from the textbook, which sounded harmless enough when Darrow read them into the record during cross-examination. Darrow also had White confirm that the state textbook commission had officially adopted the text for use in Tennessee public schools. The only clash occurred when Stewart asked White to identify the King James version of the Bible, and offered it as evidence "of what the act relates to when it says 'Bible.' " Seizing an opportunity to emphasize differences in biblical interpretation, Hays immediately objected on the grounds that dozens of differing versions of the Bible existed. "This is a criminal statute and should be strictly construed. There is nothing in the statute that shows [teachers] should be controlled in their teaching by the King James version," he declared. In Protestant East Tennessee, however, this version was the Bible—or so Judge Raulston stated in overruling the objection.[9]

Two high school students followed White to the witness stand. The first, Howard Morgan, a wide-eyed freshman from Scopes's general science course, testified that the defendant once discussed human evolution in class. "Well, did he tell you anything else that was wicked?" Darrow asked on cross-examination. Even Bryan cracked a smile as Morgan answered, "No, not that I remember."[10] Everyone laughed when the boy, who had twisted his tie under his ear and popped the top button of his shirt, assured Darrow he was not hurt by what he learned from Scopes. A sullen senior named Harry Shelton next confirmed that Scopes conducted a biology review session using Hunter's textbook. In his cross-examination, Darrow drew out that Shelton had remained a church member despite attending the class. Instruction in human evolution hardly seemed harmful to these students.

Finally, Frank Robinson took the stand. He testified that Scopes himself admitted that "any teacher in the state who was teaching Hunter's Biology was violating the law." On cross-examination, Darrow asked him about this textbook. "You were selling them, were you not?" and "You were a member of the school board?" Spectators began to laugh as they caught Darrow's drift, and broke out again when he cautioned the witness, as if advising a bootlegger, "You are not bound to answer these questions." Stewart could only joke back, "The law says teach, not sell."[11] Darrow was having too much fun at the prosecution's expense. Stewart decided against calling further witnesses. He simply stated that others were prepared to offer similar testimony and rested the state's case less than an hour after it began.

With the afternoon drawing to a close, Darrow called the defense's first witness, the zoologist Maynard M. Metcalf. Stewart interrupted to remind Darrow, "We have a rule in this state that precludes the defendant from taking the stand if he does not take the stand first." Darrow turned to the judge, "Your honor, every single word that was said against this defendant, everything was true." Scopes would not testify, Darrow declared.[12] Rather than deny what Scopes had done, the defense would seek to show that his actions did not violate the law—and this required expert testimony about the theory of evolution and the Bible. "So I sat speechless, a ringside observer at my own trial, until the end of the circus," Scopes later commented. "Darrow realized that I was not a [biology] teacher and he was afraid that if I were put on the stand I would be asked if I actually taught biology," he added. "Although I knew something of science in general, it would be quite another matter to deal exhaustively with scientific questions on the witness stand."[13]

A 57-year-old senior scientist, Metcalf represented the logical choice as the first, and perhaps only, witness for the defense. He had graduated from Oberlin College when it still had strong evangelical Protestant ties, and returned to that college after earning a doctorate in zoology from one of the nation's leading research institutions, Johns Hopkins University. Gaining a solid reputation as a researcher and teacher, Metcalf served as an officer in several professional associations. He moved on to Washington during the First World War as chief of the biology and agriculture division of the National Research Council, and afterward became a senior researcher at Johns Hopkins. Metcalf remained active in the church throughout his career and taught the college-age Sunday school class at modernist Congregational churches in Oberlin and Baltimore; now he would try to teach the court and the country about the theory of evolution and its compatibility with Christianity.

After establishing Metcalf's credentials, Darrow asked him, "Will you state what evolution is, in regard to the origin of man?" Stewart jumped up on this cue. "We except to that," he interjected. "We are excepting to everything here that pertains to evolution or to anything that tends to show that there might be or might not be a conflict between the story of divine creation and evolution."[14] The prosecution maintained that the statute outlawed any teaching about human evolution regardless of what evolution meant or whether it conflicted with the Bible. This position rendered evidence on those questions irrelevant. The defense countered that the law only barred instruction in evolution that denied the biblical account of creation, and therefore such evidence was relevant. Indeed, it constituted the defense's entire case. The judge decides questions regarding the admissability of evidence, so the jurors again left the room, less than two hours after they entered it, and remained out for the rest of the week. The judge would hear Metcalf's testimony that afternoon. The two sides then would argue the question of its admissability on Thursday. Even if the court excluded expert testimony, the defense still could submit evidence into the record for appeal.

"Then began one of the clearest, most succinct and withal most eloquent presentations of the case for the evolutionists that I have ever heard," Mencken wrote of Metcalf's testimony. "Darrow steered [Metcalf] magnificently. A word or two and he was howling down the wind. Another and he hauled up to discharge a broadside."[15] Darrow asked Metcalf to explain the theory of evolution, assess its status among scientists, and discuss its relationship to the biblical account of creation. "Evolution and the theory of evolution are fundamentally different. The fact of evolution is a thing that is perfectly and absolutely clear," the professor began. "But there are many points—theoretical points as to the methods by which evolution has been brought about—that we are not yet in possession of scientific knowledge to answer. We are in possession of scientific knowledge to answer directly and fully the question: 'Has evolution occurred?' "[16] Metcalf proceeded to relate technical evidence for evolution and affirm its universal acceptance among biologists, but never got around to the Bible before time for adjournment. The prosecutors silently listened to the detailed testimony. Raulston appeared sincerely interested. The audience thinned noticeably, however, with one departed spectator muttering, "He is about as authoritative as the evening breeze." After court adjourned for the day, Bryan affably handed Darrow a tiny wooden monkey as a memento of the case.[17] One day of cordial relations had netted substantial progress.

Thursday thrust the participants back into conflict, as nearly every lawyer in the case interjected his views on the thorny question of expert testimony. Speakers stayed pretty much on point during the morning. So weakened by the heat that he could barely speak above a whisper, William Bryan, Jr., opened for the state and, drawing on his experience as a federal prosecutor, precisely laid out the strict, nationally accepted rules that then governed the admissability of expert evidence. "This young lawyer is not the orator his father is," one observer noted. "But he seems to have a liking for matters of fact which distinguishes him from his father. He read citation after citation of dry cases with apparent pleasure."[18] As young Bryan correctly concluded to the court from these cases, if the statute simply barred teaching evolution, then "to permit an expert to testify upon this issue would be to substitute trial by experts for trial by jury, and to announce to the world your honor's belief that this jury is too stupid to determine a simple question of fact."[19]

Hays responded by stressing the defense's interpretation of the statute. "Oh, no, the law says that [a guilty teacher] must teach a theory that denies the story as stated in the Bible. Are we able to say what is stated in the Bible? Or is it a matter of words literally interpreted?" The same evidence rules cited by the younger Bryan, under Hays's view of the statute, became arguments for admitting expert testimony from scientists and theologians. Hays concluded by reminding Raulston of the broad issues at stake. "The eyes of the country, in fact of the world, are upon you here," he pleaded. "This is not a case where the sole fact at issue is whether or not Mr. Scopes taught evolution."[20]

Daytonians originally welcomed a broad test of the antievolution statute; now they shunned it. "This is a court of law, it is not a court of instruction for the mass of humanity at large," Herbert Hicks told the judge. Ben McKenzie seconded this remark in his own colorful fashion. "We have done crossed the Rubicon. Your honor has held that the act was reasonable," he proclaimed. "That never left anything on the face of the earth to determine, except as to the guilt or innocence of the defendant." Displaying a fundamentalist suspicion of academic theologians, both prosecutors also questioned the need for expert testimony to interpret scripture even under Hays's view of the statute. "Why should these experts know anything more about the Bible than some of the jurors?" Hicks asked. Amid shouts of "amen" from local spectators, McKenzie maintained that the Genesis account "is a much more reasonable story to me than that God threw a substance into the sea and said, 'In sixty thousand years, I'll make something out of you.' " When Hays challenged this, McKenzie asked him, "Do you believe the story of divine creation?" Hays retreated with the words, "That is none of your business."[21] Yet primary responsibility for answering Hays fell to the elder Bryan; Malone and Stewart would close. The hour approached noon however, and rather than risk interrupting the Commoner's oratory, Raulston adjourned early for lunch. During the extended recess, workers finally installed ceiling fans in the courtroom.

"Word that the great Bryan was to speak made the courthouse a magnet, and long before the time set for the afternoon session of the Scopes trial the crowds filled the courtroom," Philip Kinsley reported for the *Chicago Tribune*. "Out under the cottonwoods, in a much cooler situation, the greater crowds gathered to hear the story from the brazen mouths of the loud speakers. The whole town was one great sounding board of oratory."[22] No one left disappointed. Bryan was brilliant; Malone more so. Stewart stopped the show. The judge tried to avert outbursts by warning spectators at the outset, "The floor on which we are now assembled is burdened under great weight . . . so I suggest to you to be as quiet in the courthouse as you can; have no more emotion than you can avoid; especially have no applause."[23] No mere threat of physical catastrophe, however, could still the emotions unleashed that afternoon. As Darrow inserted at this point in his autobiographical account of the trial, "All in all, that was a summer for the gods!"[24]

Bryan began and ended talking about expert witnesses but in between soared into an hour-long assault against teaching evolution. "Your honor, it isn't proper to bring experts in here to try to defeat the purpose of the people of this state by trying to show that this thing that they denounce and outlaw is a beautiful thing," he began. And the people denounce it because it undermines morality, Bryan asserted. "This is that book!" he exclaimed, holding aloft Hunter's *Civic Biology*. "There is the book they were teaching your children that man was a mammal and so indistinguishable among the mammals that they leave him there with 3,499 other mammals. Including elephants!" the old Democrat charged in a joking reference to Republicans. "Talk about putting Daniel in the lion's den!" The audience hung on every word, and laughed on cue. "The Christian believes man came from above, but the evolutionist believes he must have come from below," Bryan thundered. He then quoted liberally from Darrow's statements at the Leopold–Loeb trial to argue that Darwinian teaching encouraged selfish, animalistic behavior. "Now, my friends, Mr. Darrow asked Howard Morgan, 'Did it hurt you?' " Bryan observed regarding Scopes's teaching. "Why did he not ask the boy's mother?"[25]

This rhetorical question brought Bryan back to the issue of expert witnesses. "When it comes to Bible experts, do they think that they can bring them in here to instruct members of the jury?" he asked. "The one beauty about the Word of God is, it does not take an expert to understand it." Bryan concluded to great applause, "The facts are simple, the case is plain, and if these gentlemen want to enter upon the field of education work, . . . then convene a mock court for it will deserve the title of mock court if its purpose is to banish from the hearts of people the Word of God as revealed."[26]

Malone responded with a moving, half-hour plea for freedom that had all the more impact because the audience did not see it coming. "As he rose to answer Bryan he performed the most effective act anyone could have thought of to get the audience's undivided attention: He took off his coat," Scopes later recalled. "Every eye was on him before he said a single word."[27] Malone started off quietly, half-seated on the defense table—as if humbled to follow the Great Commoner. "I defy anybody, after Mr. Bryan's speech, to believe that this is not a religious question," Malone commented near the beginning. "Oh, no, the issue is as broad as Mr. Bryan himself makes it." The volume rose as Malone recalled the Commoner's pretrial threats. "We have come here for this duel," Malone declared, "but does the opposition mean by duel that one defendant shall be strapped to a board and that they alone shall carry the sword? Is our only weapon—the witnesses who shall testify to the accuracy of our theory—is our only weapon to be taken from us?" By now he stood tall and shouted his defiance. "We

Dudley Field Malone, arms crossed, addressing crowded courtroom with WGN microphone in foreground and newsreel camera in background. Prosecutors Wallace Haggard, left, and Herbert Hicks stand immediately behind Malone. (Courtesy of Bryan College Archives)

feel we stand with science. We feel we stand with intelligence. We feel we stand with fundamental freedom in America. We are not afraid," Malone concluded. "We ask your honor to admit the evidence as a matter of correct law, as a matter of sound procedure and as a matter of justice to the defendant."[28]

"Dayton thundered her verdict at the end of the speech of Malone," one reporter observed. "Women shrieked their approval. Men, unmoved even by Darrow, could not restrain their cheers." The applause clearly eclipsed that given Bryan the hour before. An Irish policeman, on loan from Chattanooga, pounded the table so hard with his nightstick that the surface split; when another officer rushed forward to assist him in quieting the crowd, the first one shouted back, "I'm not trying to restore order. Hell, I'm cheering." The assembled press broke its customary neutral silence by giving Malone a standing ovation. Mencken hailed him with the words, "Dudley, that was the loudest speech I ever heard." Antievolution lawmaker turned press commentator J. W. Butler called it "the finest speech of the century." Of course, the address appealed to the crowd's sense of justice and avoided assaults on local religious sentiment, but it clearly stirred an emotional response. "Malone's words, read today, seem dry and uninspiring; delivered in full heat of battle . . . they were electric," Scopes wrote four decades later. "His reply to Bryan was the most dramatic event I have attended in my life."[29]

Following such flights of oratory, the task of closing for the state fell on Stewart. He returned to the solid ground of statutory interpretation to refute the need for expert testimony. The antievolution statute was clear: "Who can come here to say what is the law is not the law?" Stewart asked. "What will these scientists testify? They will say [evolution] was simply the method by which God created man. I don't care. This act says you cannot [teach it]." Stewart had not expressed strong feelings against teaching evolution before the trial, but Bryan's words led him to take the matter seriously. "We have the right to participate in scientific investigation, but, if the court please, when science strikes upon that which man's eternal hope is founded, then I say the foundation of man's civilization is about to crumble," he declared. "Shut the door to science when science sets a canker on the soul of a child." Moved by the oratory of the day, the usually reserved Stewart thrust his arms toward heaven and confessed, "They say it is a battle between religion and science. If it is, I want to serve notice now, in the name of the great God, that I am on the side of religion . . . because I want to know beyond this world that there might be an eternal happiness for me and for others." When Hays tried to interrupt by asking for a chance to prove the truth of evolution, Stewart cut him off: "That charge strikes at the very vitals of civilization and Christianity, and is not entitled to a chance."[30] The fickle crowd again erupted in sustained applause as Stewart concluded the afternoon's debate. Court adjourned until Friday morning, when the judge would rule on the state's motion to exclude expert testimony.

Dayton buzzed over the day's proceedings until late into the night. "Weary telegraphers worked until the morning hours to give the country the story of that eventful session," the *Nashville Banner* reported. "The big hour has gone into history, and all prophesies that the trial of the Rhea County school teacher would rank with that of Galileo are well nigh fulfilled."[31] The *New York Times* called it "the greatest debate on science and religion in recent years," and printed the complete text of the speeches by Bryan and Malone beginning on the front page.[32] Two sentences from those speeches stood out, and provided grist for headline writers across the country. "'They Call Us Bigots When We Refuse to Throw Away Our Bibles,' Asserts Bryan," proclaimed the lead article on the *Chicago Tribune* wire.[33] "We say 'Keep Your Bible,'" read the headlined response from Malone, "but keep it where it belongs, in the world of your conscience . . . and do not try to tell an intelligent world and the intelligence of this country that these books written by men who knew none of the accepted fundamental facts of science can be put into a course of science."[34] Each side had effectively communicated its message and celebrated its success, but neither offered any basis for compromise in this contest for America's heart and mind.

Court met for less than an hour on Friday, just enough time for the judge to rule out expert testimony but not quite long enough for him to cite Darrow for contempt. The two developments were related. From the start, Raulston sounded uncharacteristically defensive. He clearly wanted to hear the experts but felt pressure from state leaders who, fearing that such testimony would heap further ridicule on Tennessee and its law, pointedly had declared that the trial should be brief. The judge stumbled badly in reading his ruling, which adopted the prosecution's position precisely; it was a plausible position—even the otherwise critical *New York Times* grudgingly endorsed it, as did many respondents to an informal survey of East Tennessee lawyers conducted by the prodefense *Chattanooga Times.* Nevertheless, Hays and Darrow immediately confronted Raulston, hoping to expose judicial bias. They had little to lose at this point. Hays objected to the ruling in a contemptuous manner. Stewart took exception to Hays's tone: "I think it is a reflection upon the court." Raulston brushed it aside with the words, "Well, it don't hurt this Court." Darrow picked up the assault. "There is no danger of it hurting us," he mocked. "The state of Tennessee *don't* rule the world yet."[35]

Hays asked about submitting expert testimony to the court for the purpose of creating a record for appellate review. Raulston offered to let the experts either submit sworn affidavits or summarize their testimony for the court reporter. When Hays pressed for live testimony by the experts, Bryan interrupted him. "If these witnesses are allowed to testify," he queried, "I presume they would be subject to cross-examination?" Bryan struck a nerve. Hays later explained the dilemma: "Cross-examination would have shown that the scientists, while religious men—for we chose only that kind—still did not believe in the virgin birth and other miracles." Such testimony would undermine the ACLU's broader efforts. "It was felt by us that if the cause of free education was ever to be won, it would need the support of millions of intelligent churchgoing people who didn't question theological miracles," Hays noted. Immediately Darrow struck back in court: "They have no more right to cross-examine—" Raulston saw his chance to turn the tables on his tormentor. "Colonel, what is the purpose of cross-examination? [I]sn't it an effort to ascertain the truth?" Darrow hunched up his shoulders to his ears and stared at the judge. "Has there ever been any effort to ascertain the truth in this case?" he shot back. The defense could submit written affidavits or read prepared statements into the record, the court ruled, but the prosecution could cross-examine any witnesses put on the stand.[36]

Darrow requested the rest of the day to compose the witness statements. When Raulston questioned the need for so much time, Darrow exploded. "I do not understand why every request of the state and every suggestion of the prosection should meet with an endless waste of time, and a bare suggestion of anything on our part should be immediately over-ruled," he shouted. "I hope you do not mean to reflect upon the court?" Raulston demanded. Darrow tugged on his suspenders and carefully weighed his response. "Well, your honor has the right to *hope,*" came the answer, with menacing emphasis on the final word. "I have the right to do something else, perhaps," the judge declared, but agreed to recess court until Monday so that the defense witnesses could prepare their statements.[37]

"All that remains of the great cause of the State of Tennessee against the infidel Scopes is the final business of bumping off the defendant," Mencken wrote in his final report from Dayton. "There may be some legal jousting on Monday and some gaudy oratory on Tuesday, but the main battle is over, with Genesis completely triumphant."[38] That was the general consensus on Saturday, as Mencken and dozens of other crack journalists departed Dayton just as Darrow plotted his comeback. "To the newspaper men the trial drags along slowly, tiresomely, bitterly," one of them commented that day. "The reporters are all sick of the over-cooked food, the upsetting water, the jungle heat, and the exhausting, Herculean work, recording a play of emotions ranging from hateful words that sting like bullets to bowing and scraping court manners smacking of ill-concealed deceit. Everyone's nerves are racked to shreds."[39] Of course, Mencken had additional incentive to leave early. Local officials, angered by his slurs of Dayton and its inhabitants, warned him to go before townspeople forced him out. "I hope nobody lays hands on him," Chief Commissioner A. P. Haggard told the press. "I stopped them once, but I may not be there to dissuade them if it occurs to them again."[40] Yet Mencken surely would have run this risk had he known what Darrow planned for Monday.

"I'm going to put a Bible expert on the stand 'bout day after tomorrow," Charles Francis Potter later recalled Darrow confiding to him on Saturday. "A greater expert than you—greatest in the world—he thinks." Potter understood immediately, and called it a master stroke. "Never mind the master stuff," Darrow replied, "and don't talk so loud. Too many reporters round here."[41] While a dozen defense experts labored over their statements at the Mansion, Darrow quietly prepared to call Bryan to the witness stand. Darrow rehearsed the interrogation on Sunday night with the Harvard geologist, Kirtley Mather, playing the Commoner's role, using the same type of questions he asked Bryan two years earlier in a public letter to the *Chicago Tribune.* By Sunday, the press began to sense something was afoot; the *Nashville Banner* reported, "Rumors go about that the defense is preparing to spring a coup d'etat."[42]

Oblivious to Darrow's scheme, prosecutors basked in their apparent victory. Stewart pronounced the judge's ruling "a glorious victory." William Bryan, Jr., headed home to California confident in the trial's outcome. His father put the finishing touches on his closing arguments, which he promised would "be something brand new," and began looking beyond the trial. The elder Bryan talked about carrying the fight against teaching evolution to seven other state legislatures during the next two years and, on Saturday, issued a long written statement hailing the trial's impact. "We are making progress. The Tennessee case has uncovered the conspiracy against Biblical Christianity," Bryan wrote, and "unmasked" the "cruel doctrine" of natural selection that robs civilization of pity and mercy. He denounced Darrow as the chief conspirator. "He protested against opening court with prayer and has lost no opportunity to slur the intellect of those who believe in Orthodox Christianity," Bryan charged. "Mr. Darrow's connection with this case and his conduct during this case ought to inform the Christian world of the real animus that is back of those who [promote the teaching of evolution]." On Sunday, Bryan made similar comments while preaching at a combined open-air service of all churches in Pikeville, a small town fifteen miles west of Dayton.[43]

Darrow's effort to suppress court prayer evoked strong negative reactions and stirred a nation that still clung to displays of civil religion. Newspaper editorials throughout the country criticized Darrow for the move. Governor Peay broke his silence on the trial with the public comment, "It is a poor cause that runs from prayer." The Florida philanthropist George F. Washburn was so enthusiastic about developments in the trial that he pledged $10,000 toward Superintendent White's idea of founding Bryan University. "This fight in Dayton is 'the beginning of the battle that will encircle the world,'" Washburn wrote to White. "This is a psychological moment to establish a Fundamentalist university." As if to overwhelm this offer and swamp its potential impact, John D. Rockefeller, Jr., gave $1 million to Shailer Mathews's University of Chicago Divinity School on the same day.[44]

That Saturday, Darrow responded to Bryan's written statement with one of his own. He backtracked on the matter of court prayer by suggesting that he only "objected to it because of the peculiar situation [of this case]" but held his ground on evolutionary naturalism as a basis for morality. "I hope this [philosophy] has made me more understanding of my fellowmen and kindlier and more charitable to them," Darrow wrote. He returned to the issue of ethics and evolution on Sunday in the course of delivering a lecture on Tolstoy to a Chattanooga Jewish group. Over the weekend, Darrow took time off to witness the ecstatic worship of Pentecostals encamped near Dayton. "They are better than Bryan," he laughed to reporters. Darrow got in his best shot after Bryan told the press that, although the law required excluding expert testimony from the trial, "personally" he wanted to hear it. "Bryan is willing to express his opinions on science and religion where his statements will not be questioned," Darrow replied to reporters in a statement clearly calculated to bait the Commoner, "but Bryan has not dared to test his views in open court under oath."[45]

As Darrow verbally jousted with Bryan, Hays assumed responsibility for overseeing the preparation of witness statements. The Mansion became the locus of activity. "Here the affidavits of the scientists are being whipped into final shape," one observer noted. "They will approach the problem from almost as many angles as there are fields of natural science. Zoology, animal husbandry, agronomy, geology, botany, anthropology, every one will be presented. Each will give the testimony that evolution is a fact, and that a well rounded education cannot well do without it."[46] In all, eight scientists dictated a total of more than 60,000 words of testimony to stenographers, who typed formal statements for the court and duplicated copies for the press. In addition, four religion experts summarized their testimony for Hays to read in court. After completing their written statements over the course of the weekend, most of the experts packed up and left town.

With Bryan preaching in Pikeville, Darrow lecturing in Chattanooga, and the defense witnesses closeted in the Mansion, Dayton returned to near normal that second weekend. The *Nashville Tennessean* described the scene as "dead calm."[47] Many reporters had departed for good; most of the rest spent Saturday and Sunday in Chattanooga. Downtown concessioners complained of losing money. A handful of visitors turned out for a high school band concert on the courthouse lawn. Scopes went swimming in the mountains. One or two days of oratory remained, court observers predicted, and the town would slip back into obscurity. "Dayton," the *Nashville Banner* reported, "looks forward to the coming of Monday with the same anticipation as a man who has just eaten a hardy dinner contemplates breakfast in the morning. It does not seem possible that anything can transpire to make the trial of John T. Scopes interesting again."[48] As it turned out, Darrow had more imagination than Dayton.

The day started out as everyone expected. The crowd formed early in anticipation of hearing closing arguments and filled every seat in the gallery by 8:30 A.M. Police then closed the main entrance, but people continued to slip in through a side door. Judge Raulston could barely squeeze through the aisles when he arrived shortly after 9:00. A fundamentalist pastor opened the proceedings with a confessional prayer aimed directly at the defense: "Thou has been constantly seeking to invite us to contemplate higher and better and richer creations of Thine, and sometimes we have been stupid enough to match our human minds with revelations of the infinite and eternal."[49] Before anyone else could speak, the judge began reading from a prepared statement citing Darrow in contempt of court for his remarks on Friday, and ordering him to appear on Tuesday for sentencing. Over the weekend, leaders of the Tennessee bench and bar had urged action on the matter. Darrow expected it, and did not protest. Chattanooga attorney Frank Spurlock immediately offered to post Darrow's bond, and the trial got back on track.

Defense counsel turned to the question of how to submit its expert testimony into the appellate record. Typically, local attorneys dictated a summary of excluded evidence to a court reporter or provided written statements from barred witnesses, but here the defense still hoped to educate the public about evolution. Hays asked to read the experts' statements in open court with the jury excused. Stewart insisted that they simply be added to the written record. The wrangling continued for nearly an hour, until Bryan suggested that the prosecution should have an opportunity to respond to any oral presentations by the defense. Darrow quickly offered a compromise: Submit the written statements but allow Hays to read selected excerpts in court. Raulston agreed, and gave Hays one hour. He took over two, and did not finish until after lunch.

The excerpts read by Hays made no appreciable impact within the courtroom. They laid out the case for evolution in great detail but were ill-suited for recitation on a hot summer day. In all, eight scientists provided written statements on evolution: the anthropologist Fay Cooper Cole, the psychologist Charles Hubbard Judd, and the zoologist Horatio H. Newman, all from the University of Chicago; the University of Missouri zoologist Winterton C. Curtis; the Rutgers agronomist Jacob G. Lipman; the Harvard geologist Kirtley F. Mather; the Johns Hopkins zoologist Maynard M. Metcalf; and the state geologist Hubert A. Nelson of Tennessee, who the defense added to its witness list after Bryan began criticizing its reliance on out-of-state experts. In their statements, Curtis, Mather, and Metcalf also sought to reconcile the theory of evolution with the biblical account of creation, as did testimony submitted from four religion experts: Shailer Mathews; Herman Rosenwasser (a Hebrew Bible scholar who appeared in Dayton without invitation but quickly impressed defense counsel); and two Tennessee modernists, the Methodist minister Herbert E. Murkett of Chattanooga and the Episcopal priest Walter C. Whitaker of Knoxville.[50] The jury heard none of it.

When court reconvened following lunch, Darrow interrupted the presentation of testimony to apologize for his comments on Friday. Townspeople had treated him courteously, Darrow cooed, and he should not have responded to the court as he did. "One thing slipped out after another," Darrow explained, "and I want to apologize to the court for it." Rising to his feet, Raulston dismissed the contempt citation with words that amazed the defense. After discussing the honor of Tennessee, he recited from memory a long religious poem about forgiveness and accepted Darrow's apology in the name of Christ. "We forgive him," the judge said of Darrow in a voice shaking with emotion, "and we command him to go back home and live in his heart the words of the Man who said: 'If you thirst come unto Me and I will give thee life.' "[51] Christianity represented more than civil religion in this court.

Raulston shifted his concern to the crowd that overflowed the courtroom. A rumor spread that cracks had appeared in the ceiling downstairs. Scopes thought it was simply "the man-killing heat."[52] Most likely, the judge thought that little remained except closing arguments and wanted to give everyone an opportunity to see them. Whatever the reason, he moved the proceedings to the speaker's platform on the courthouse lawn. "It was a striking scene. Judge Raulston sat at a little wooden table in the center, with the States attorneys at his left and the defense at his right," wrote one observer. "In front was a sea of upturned faces, waiting for what they presumed would be an ordinary argument, faces which became eager when Mr. Darrow announced that he would call Mr. Bryan as a witness for the defense."[53]

In fact it was Hays who summonsed Bryan, but only after finishing the witness statements. Next, Darrow objected to a Read Your Bible banner hanging on the courthouse near the makeshift jury box. Bryan conceded that the sign might appear prejudicial and it was taken down. Then, with the jury still excused, Hays called Bryan as the defense's final expert on the Bible and the Commoner again proved cooperative. Up to this point, Stewart had masterfully confined the proceedings and, with help from a friendly judge, controlled his wily opponents. Indeed, Governor Peay had just wired the young prosecutor, "You are handling the case like a veteran and I am proud of you."[54] Yet Stewart could not control his impetuous co-counsel and the judge seemed eager to hear the Peerless Leader defend the faith. "All the lawyers leaped to their feet at once," Scopes recalled.[55] Ben McKenzie objected. Stewart seethed with anger. Bryan consented solely on condition that he later get to interrogate Darrow, Malone, and Hays. "All three at once?" Darrow asked. As Bryan explained early in his testimony, "They did not come here to try this case, . . . They came here to try revealed religion. I am here to defend it, and they can ask me any questions they please."[56] Darrow did just that.

The crowd swelled as word of the encounter spread. From the 500 persons who evacuated the courtroom, the number rose to an estimated 3,000 people sprawled across the lawn—nearly twice the town's normal population. "The spectators, however, instead of being only men, were men, women, and children, and among them here and there a negro," the *New York Times* reported. "Small boys went through the crowd selling bottled pop. Most of the men wore hats and smoked."[57] The *Nashville Banner* added, "Then began an examination which has few, if any, parallels in court history. In reality, it was a debate between Darrow and Bryan on Biblical history, on agnosticism and belief in revealed religion."[58] Darrow posed the well-worn questions of the village skeptic, much like his father would have asked in Kinsman, Ohio, fifty years earlier: Did Jonah live inside a whale for three days? How could Joshua lengthen the day by making the sun (rather than the earth) stand still? Where did Cain get his wife? In a narrow sense, as Stewart persistently complained, Darrow's questions had nothing to do with the case because they never inquired about human evolution. In a broad sense, as Hays repeatedly countered, they had everything to do with it because they challenged biblical literalism. Best of all for Darrow, no good answers existed to them. They compelled Bryan "to choose between his crude beliefs and the common intelligence of modern times," Darrow later observed, or to admit ignorance.[59] Bryan tried all three approaches at different times during the afternoon, without appreciable success.

Darrow questioned Bryan as a hostile witness, peppering him with queries and giving him little chance for explanation. At times it became like a firing line:

> "You claim that everything in the Bible should be literally interpreted?"
>
> "I believe everything in the Bible should be accepted as it is given there; some of the Bible is given illustratively. . . . "
>
> "But when you read that . . . the whale swallowed Jonah . . . how do you literally interpret that?"
>
> ". . . I believe in a God who can make a whale and can make a man and make both of them do what he pleases. . . . "
>
> "But do you believe he made them—that he make such a fish and it was big enough to swallow Jonah?"
>
> "Yes sir. Let me add: One miracle is just as easy to believe as another."
>
> "It is for me . . . just as hard."
>
> "It is hard to believe for you, but easy for me. . . . When you get beyond what man can do, you get within the realm of miracles; and it is just as easy to believe the miracle of Jonah as any other miracle in the Bible."[60]

Such affirmations undercut the appeal of fundamentalism. On the stump, Bryan effectively championed the cause of biblical faith by addressing the great questions of life: The special creation of humans in God's image gave purpose to every person and the bodily resurrection of Christ gave hope to believers for eternal life. Yet Darrow did not inquire about these grand miracles. For many Americans, laudable simple faith became laughable crude belief when applied to Jonah's whale, Noah's Flood, and Adam's rib. Yet the Commoner acknowledged accepting each of these biblical miracles on faith and professed that all miracles were equally easy to believe.

Bryan fared little better when he tried to rationalize two of the biblical passages raised by Darrow. In an apparent concession to modern astronomy, Bryan suggested that God extended the day for Joshua by stopping the earth rather than the sun; similarly, in line with nineteenth-century evangelical scholarship, Bryan affirmed his understanding that in Genesis, days of creation represented periods of time, which led to the following exchange:

> "Have you any idea of the length of these periods?"
>
> "No; I don't."
>
> "Do you think the sun was made on the fourth day?"
>
> "Yes."
>
> "And they had evening and morning without the sun?"
>
> "I am simply saying it is a period."
>
> "They had evening and morning for four periods without the sun, do you think?"
>
> "I believe in creation as there told, and if I am not able to explain it I will accept it."[61]

Although Bryan had not ventured far beyond the bounds of biblical literalism, the defense made the most of it. "Bryan had conceded that he interpreted the Bible," Hays gloated. "He must have agreed that others have the same right."[62] Furthermore, Scopes observed, "The Bible literalists who came to cheer Bryan were surprised, ill content, and disappointed that Bryan gave ground."[63]

As Darrow pushed various lines of questioning, increasingly Bryan came to admit that he simply did not know the answers. He had no idea what would happen to the earth if it stopped moving, or about the antiquity of human civilization, or even about the age of the earth. "Did you ever discover where Cain got his wife?" Darrow asked. "No sir; I leave the agnostics to hunt for her," came the bittersweet reply.[64] "Mr. Bryan's complete lack of interest in many of the things closely connected with such religious questions as he had been supporting for many years was strikingly shown again and again by Mr. Darrow," the *New York Times* reported.[65] Stewart tried to end the two-hour interrogation at least a dozen times, but Bryan refused to step down. "I am simply trying to protect the word of God against the greatest atheist or agnostic in the United States," he shouted, pounding his fist in rage. "I want the papers to know I am not afraid to get on the stand in front of him and let him do his worst."[66] The crowd cheered this outburst and every counterthrust attempted by the Commoner. Darrow received little applause but inflicted the most jabs. "The only purpose Mr. Darrow has is to slur the Bible, but I will answer his questions," Bryan exclaimed at the end. "I object to your statement," Darrow shouted back, both men now standing and shaking their fists at each other. "I am examining your fool ideas that no intelligent Christian on earth believes."[67] Raulston finally had heard enough and abruptly adjourned court for the day.

Darrow's supporters rushed forward to congratulate their hero. Bryan was left virtually alone with his thoughts. Reporters rushed out to transmit the news. "Men who have written descriptions of great battles," one journalist commented, "were overwhelmed with their responsibility to give their papers an account of the two sessions of court Monday in a light which will depict truly its immensity."[68] Newspapers throughout the country printed the complete transcript. The Memphis *Commercial Appeal* concluded: "It was not a contest. Consequently there was no victory. Darrow succeeded in showing that Bryan knows little about the science of the world. Bryan succeeded in bearing witness bravely to the faith which he believes transcends all the learning of men."[69] Most papers were not so kind to Bryan. That night, Stewart delivered the word to Bryan: he should neither resume his testimony nor call defense counsel to the stand. At first Bryan protested, but Stewart stated that if Bryan demanded to go forward either the judge would forbid it or the state would dismiss the case. Darrow shortly wrote to Mencken about the examination of Bryan, "I made up my mind to show the country what an ignoramus he was and I succeeded."[70]

A light rain fell in Dayton on Tuesday morning, forcing the trial back inside and cutting the number of spectators. The judge started the proceedings a few minutes early, before Darrow and Stewart arrived. As counsel for both sides hurried to their places, Raulston took it upon himself to bar further examination of Bryan and to order the Commoner's prior testimony expunged from the record. "I feel that the testimony of Mr. Bryan can shed no light upon any issue that will be pending before the higher court," the judge ruled; "the issue now is whether or not Mr. Scopes taught that man descended from a lower order of animals."[71]

With this ruling, Darrow called it quits. "We have no witnesses to offer, no proof to offer on the issues that the court has laid down here," he declared. "I think to save time we will ask the court to bring in the jury and instruct the jury to find the defendant guilty." Stewart immediately agreed: "We are pleased to accept the suggestion of Mr. Darrow." This final ploy by the defense deprived Bryan of the chance to deliver his closing argument, and also averted the risk of a hung jury, which would have frustrated defense plans to challenge the constitutionality of the antievolution statute in a higher court. Bryan accepted the inevitable. "I shall have to trust to the justness of the press, which reported what was said yesterday, to report what I will say, not to the court, but to the press in answer," he told the court, "and I shall also avail myself of the opportunity to give to the press, not to the court, the questions that I would have asked had I been permitted to call the attorneys on the other side."[72]

Stewart, Darrow, and Raulston agreed on the terms of the judge's charge to the jury, and jurors finally reentered the courtroom. After expecting front-row seats for the entire proceedings, they had heard only two hours of testimony against Scopes and none of the memorable speeches. Based on what the court had permitted them to hear, Darrow told jurors, "We cannot even explain to you that we think you should return a verdict of not guilty. We do not see how you could. We do not ask it." Stewart merely added, "What Mr. Darrow wanted to say to you was that he wanted you to find his client guilty, but did not want to be in the position of pleading guilty, because it would destroy his rights in the appellate court."[73] At least one farmer-juror welcomed the trial's end. "The peach crop will soon be coming in," he commented to a reporter.[74]

The jury received the case shortly before noon and returned its verdict nine minutes later. They spent most of this time getting in and out of the crowded courtroom. "The jurors didn't even sit down to think it over," one observer noted, "but stood huddled together in the hallway of the courthouse for the brief interval."[75] As the court waited, Bryan turned to Malone, "I am not a gambling man, but if I was I would bet that the verdict is guilty." Malone laughed, "That is my bet too. I think we're beaten."[76] In reality, both men felt like victors that day. The only point of concern involved the jury's decision to let the judge impose the minimum $100 fine. State law required that the jury fix the amount of the penalty, Stewart observed. Raulston assured him that local practice in misdemeanor cases allowed the judge to impose the minimum on persons adjudged guilty, and Darrow agreed to that procedure—a decision he would deeply regret.

Only a few speeches remained. Scopes spoke briefly at the time of sentencing—his first words to the court. Prompted by Neal, the defendant called the antievolution statute unjust and pledged to continue fighting it in the name of academic freedom. Counsel took turns thanking the court and community. Representatives from the press and state bar added cordial comments. In their farewell remarks, Bryan and Darrow tried to explain the widespread interest in the trial. The Commoner called the matter a little case raising a great cause, and asserted

Clarence Darrow addressing Scopes jury in front of packed courtroom. (Courtesy of Bryan College Archives)

that "causes stir the world." Darrow, in contrast, blamed everything on the religious nature of the prosecution. "I think this case will be remembered because it is the first case of this sort since we stopped trying people in America for witchcraft," he claimed. "We have done our best to turn the tide . . . of testing every fact in science by a religious doctrine."[77] Raulston expressed his satisfaction with the trial and, after a local minister delivered a benediction, adjourned the court in time for lunch. As the crowd surged forward to hail both Darrow and Bryan, the Scopes trial passed into history and its legend took on a life of its own.

PART III

. . . and After

— CHAPTER EIGHT —

THE END OF AN ERA

THE SCOPES TRIAL did not end the antievolution crusade. How could it? Scopes had lost and the law was upheld. Darrow embarrassed Bryan on the witness stand, but the Commoner was an experienced politician accustomed to rallying from defeat. More critically, he was an innate optimist. "If Bryan left the Scopes trial 'an exhausted and broken man,' as one writer has recently maintained," Bryan biographer Lawrence W. Levine observed, "he did a masterly job of concealing it during the five days of life remaining to him."[1]

Bryan took the offense immediately after the trial ended. Only hours after court adjourned, Bryan released a series of curt questions to defense attorneys about their beliefs in God, biblical truth, Christ, miracles, and life after death. Darrow replied within the hour by tersely affirming his agnosticism on every point, concluding with his succinct answer as to the question of immortality: "I have been searching for proof of this all my life, with the same desire to find it that is incident to every living thing, and I have never found any evidence on the subject."[2]

The following day, Bryan launched his effort to cast events at Dayton in a favorable light. "The issue is so large that individuals and locations are relatively unimportant," he asserted in a press statement. "Is the Bible true is the question raised by the Tennessee law, and that question is answered in the affirmative as far as this trial can answer it." He identified Darrow's contemptuous behavior in court as "exhibit A" in the moral case against life without faith in God. Darrow replied in kind. "Mr. Bryan's convulsions seem due to the fact that I placed him upon the witness stand," he told reporters later in the day. "Of course, I cannot help having some pity for Mr. Bryan for being obligated to show his ignorance by simple and competent questions asked him on the witness stand."[3] By nightfall, Bryan was defending his intelligence before assembled journalists and claiming that Darrow unfairly took advantage of his lack of technical knowledge in science. In an apparent attempt to contrast his simple answers with Darrow's crafty questions, Bryan declared, "Evolution overestimates the influence of the mind on life and underestimates the influence of the heart."[4]

During the next two days, Bryan remained in Dayton revising his unused closing argument into a fiery stump speech. This 15,000-word address leveled four specific "indictments" against the theory of evolution. First, the theory contradicted the biblical account of creation. Second, its survival-of-the-fittest explanation for human development destroyed both faith in God, as exemplified by Darwin's agnosticism, and love of others, as shown by Nietzsche's philosophy. Third, the study of evolution diverted attention from spiritually and socially useful pursuits. Finally, its deterministic view of life undermined efforts to reform society. "Let us, then, hear the conclusion of the whole matter. Science is a magnificent material force, but it is not a teacher of morals. It can perfect machinery, but it adds no moral restraints to protect society from the misuse of the machine," Bryan wrote near the end. "It not only fails to supply the spiritual element needed but some of its unproved hypotheses rob [society of its moral] compass" and thus "endangers" humanity.[5]

Bryan planned to deliver this speech to audiences across the country during the coming months, counting on residual interest in the Scopes trial to draw crowds. On Friday, July 24, Bryan drove to Chattanooga, where he arranged with the *Chattanooga News* editor George Fort Milton to publish the address. Milton bitterly opposed the antievolution law, but as a lifelong Bryan Democrat he denounced Darrow's interrogation of his Peerless Leader as "a thing of immense cruelty" and predicted that a popular backlash against it could further entrench the statute.[6] On Saturday, Bryan drove from Chattanooga to Tom Stewart's home town of Winchester, where he fulfilled a promise to the prosecutor by delivering the new address. He stopped en route in tiny Jasper, where more than 2,000 people turned out to hear him rehearse a portion of the speech in an open-air forum. An even larger audience assembled in Winchester for the full oration. The historian Ray Ginger attributed this feverish activity to "desperation," but failed to note that Bryan customarily campaigned at an even more rapid pace. "If I should die tomorrow," he reportedly told a journalist in Winchester, "I believe that on the basis of the accomplishments of the last few weeks I could truthfully say, well done."[7] The Scopes trial clearly upset Bryan, but it hardly drove him to despair. Returning to Chattanooga for a night's work on galley proofs of the address, he talked of an expanded crusade against the teaching of evolution.

These plans worried Bryan's wife. "Mother was greatly opposed to father's activities in assisting the passage of the anti-evolution laws in several States," their daughter Grace later confided in a private letter. "Mother did all she could to prevent father from taking part in the Scopes trial."[8] Now, on their drive through the Tennessee countryside, Mary Baird Bryan expressed her fears that the antievolution crusade would cross the line between a narrow effort by taxpayers to control public education and a broad assault on individual freedom of speech and belief. She recorded their exchange in an account published later that year: "'Well, Mama, I have not made that mistake yet, have I?' and I replied, 'You are all right so far, but will you be able to keep to this narrow path?' With a happy smile, he said, 'I think I can.' 'But,' I said, 'can you control your followers?' and more gravely he said, 'I think I can.'"[9] He never got that chance. After returning to Dayton for Sunday services at the southern Methodist church, where he offered the morning prayers, Bryan died in his sleep during his afternoon nap. The final words of his last speech, lifted by him from his favorite hymn to dramatize the promised results of eliminating public school instruction about human evolution, seemed a fitting eulogy: "Faith of our fathers—holy faith/ We will be true to thee till death!"[10]

Word of Bryan's death reached Darrow that afternoon as he vacationed in the Smoky Mountains. "People down here believe that Bryan died of a broken heart because of your questioning," a journalist commented. Darrow reportedly shrugged his shoulders and replied, with a reference to the Commoner's legendary appetite that would thereafter color the popular image of Bryan at Dayton, "Broken heart nothing; he died of a busted belly." Back in Baltimore, Mencken characteristically joked, "God aimed at Darrow, missed, and hit Bryan instead," but privately he reportedly gloated, "We killed the son-of-a-bitch!"[11]

Darrow had not been idle during the days immediately following the trial. In addition to responding to Bryan's sporadic statements, he had given a public lecture in Knoxville and attended a farewell party thrown by departing journalists. Hays later wrote about local high school students who attended this party and recalled that Darrow "danced with and even smoked cigarettes with them." Hays and Darrow hoped to see the youths of Tennessee turn from their parents' repressive ways. "Smoking, dancing, free association between girls and boys, games and movies on Sunday had been their issues at home," Hays added. "Here [we] were champions indeed."[12] No such generation gap developed with regard to teaching evolution in Tennessee, however; the Scopes defense attracted its share of senior supporters, such as Neal, and the most enthusiastic prosecutors were under 30 years old. Darrow's prediction that the rising generation soon would repeal the antievolution statute failed to come true. Indeed, after interviewing local students about Scopes, one journalist reported as a typical response: "I like him, but I don't believe I came from a monkey."[13] Few Tennesseans—of any age—believed otherwise.

Dayton quickly returned to normal following the trial, with few visitors other than the Bryans and the Darrows remaining by the week's end. "Dayton has benefited, physically and mentally, by the 'evolution trial,'" trial promoter Fred Robinson boasted to a departing reporter, citing the refurbished courthouse and intellectual stimulation. Yet few perceived a lasting change. "Every indication is that Dayton is back where it was before the trial began, a sleepy little town among the hills," the *Nashville Banner* observed at midweek. Even Scopes shook the dust of Dayton from his feet before the week ended.[14]

Scopes no longer felt at home in Dayton. Following the verdict, the local school board offered to renew his teaching contract for another year provided that he comply with the antievolution law, but Scopes already had his sights fixed on graduate school. "Shortly before the trial was over, Kirtley Mather of Harvard and Watson Davis of the Science

Service had notified me of the scholarship fund that expert witnesses were arranging for my graduate study in whatever field I chose," Scopes later explained. "One of my most valuable windfalls at Dayton had been listening to and associating with the distinguished scientists who stayed at the Mansion. They had broadened my view of the world." By the time Bryan died, Scopes already had left for the University of Kentucky to inquire about entering law school that fall. He ultimately settled on studying geology at the University of Chicago and became a petroleum engineer. In doing so, he passed up offers to capitalize on his fame through paid appearances on the lecture and vaudeville circuits and in movies. "I knew I would not live happily in a spotlight," Scopes concluded. "The best thing to do, I was beginning to realize, was to change my life and seek anonymity."[15]

Robinson and Scopes were not alone in their efforts to determine whether the trial had any lasting potiential benefits. In the trial's immediate aftermath, both sides found reason to celebrate. The prosecution claimed a legal victory; the defense a moral one. "Each side withdrew at the end of the struggle satisfied that it had unmasked the absurd pretensions of the other," the veteran *New York Times* reporter Russell D. Owen concluded.[16] On the day of the verdict, for example, Herbert Hicks boasted in a letter to his brother Ira, "We gave the atheist Jew Arthur Garfield Hays, the agnostic Clarence Darrow, and the ostracized Catholic Dudley Field Malone, a sound licking although the papers are prejudiced against us and may not say so." His alleged victims thought nothing of the kind. From the stage of New York's *Ziegfeld Follies* two days later, Malone declared the trial a "victorious defeat" that would help assure that "future generations will know the truth." Hays reiterated this point in an essay for *The Nation*, claiming, "It is possible that laws of this kind will hereafter meet with the opposition of an aroused public opinion."[17]

Defense claims of victory clearly relied on popular reactions to the trial rather than solely on what transpired within the courtroom—and surely Malone and Hays hoped their confident claims would promote the desired response—but others closely associated with the trial tended to see momentum shifting in Bryan's favor. Mencken's final report from Dayton concluded by warning that "Neanderthal man is organizing in these forlorn backwaters of the land, led by a fanatic, rid of sense and devoid of conscience. . . . There are other States that had better look to their arsenals before the Hun is at their gates." Talking to reporters in New York on his return from the trial, Charles Francis Potter forecast a national epidemic of antievolution lawmaking following the Scopes verdict. Asked about Potter's comments, John Roach Straton readily agreed, boasting that southern states would be the first to enact such laws, followed by western ones, and that the movement then would sweep the North and West.[18]

Tennessee newspapers generally offered a more equivocal initial assessment of the trial and its impact. Observing that the prosecution upheld the law in court even as the defense promoted its cause with the general public, a *Nashville Banner* correspondent concluded, "To call it a draw would be incorrect. The state and the defense each won a clear-cut and decisive victory."[19] The dueling commentators for the *Chattanooga Times* issued a split decision, the fundamentalist asserting that "Darrow, the agnostic, and his crowd . . . met their match in the grand old Commoner," while the modernist declared Bryan "shorn of strength" by Darrow's interrogation.[20] Even the proprosecution Memphis *Commercial Appeal* did not declare a single winner after Bryan's tortured testimony. "We saw an attempted duel between science and religion at Dayton," its editor observed, but he concluded that both sides lost ground.[21]

Similarly, the nation's press initially saw little of lasting significance in the trial beyond its having exposed Bryan's empty head and Darrow's mean spirit. After the verdict, a feature article in *The New Republic* denounced the trial as "a trivial thing full of humbuggery and hypocrisy."[22] Next-day editorials in both the *New York Times* and the *Chicago Tribune* predicted that the fight for and against teaching evolution would continue unabated.[23] The *Literary Digest* for the week summed up consensus among the press: "The trial at Dayton is no more than an opening skirmish, 'a clash of picket hosts that can not be decisive,' remarks the New York *Evening Post;* and other papers and commentators agree that it may mark the beginning of a great fight between the Fundamentalists and the Modernists."[24]

The same newspapers and magazines that dismissed the results as indecisive had given the trial unprecedented coverage only a week earlier; journalists wired two million words from Dayton during the trial, including more from America to Europe than just about any prior news event. Yet the *New York Times*—which used as many as five telegraph wires at a time to carry reports from Dayton—observed afterward that the trial's abrupt end saved "the public from having its ears bethumped with millions more of irrelevant words."[25] So long as the trial lasted, events at Dayton dominated the news, having received front-page coverage across the nation for a fortnight; as soon as Scopes lost, the story no longer was considered newsworthy and resurfaced only briefly when Bryan died. The Tennessee trial was simply the latest thrill during the kaleidoscopic Roaring Twenties. The great southern sociologist Howard W. Odum tracked press coverage at the time and counted that "some 2,310 daily newspapers in this country" covered the Scopes trial, and found "no periodical of any sort, agricultural or trade as well, which has ignored the subject."[26]

The timing of Bryan's death caused some to reassess the trial's potential significance. "Nothing could be more dramatic in time or in manner than the death of William Jennings Bryan, following so soon upon his appearance in the Dayton court room," Walter Lippmann wrote in the *New York World*. "His death at this time will weight his words at Dayton with the solemnity of a parting message and strengthen their effect upon his fellow citizens."[27] Although Bryan reportedly died of apoplexy, people generally assumed that the stress of the trial precipitated the attack. Many blamed Darrow personally. "I could sense an opinion forming that Bryan was a martyr who had died defending the Grand Old Fundamental Religion," Scopes commented, after a brief return visit to Dayton the following week. "Soon afterward there was a rumor about town that 'that old devil Darrow' had killed Bryan with his inquisition." In early August, George Rappleyea reported to ACLU officials in New York, "The death of Bryan swept away any victory we might have gained before the people of Tennessee; I am the only modernist that now remains in Dayton." Governor Peay made it official with a proclamation declaring that Bryan died "a martyr to the faith of our fathers" and announcing a state holiday to mark the funeral.[28] After studiously ignoring the Scopes trial for weeks, Peay began taking a keen interest in winning the appeal—an apparent reaction to the defense's treatment of Bryan and Tennessee.

The Commoner's funeral became a national event. Crowds lined the tracks as a special Pullman car carried the body to Washington for burial at Arlington National Cemetery. Thousands filed by the open coffin, first in Dayton, then in several major cities along the train route, and finally in the nation's capital. Flags flew at half-mast. Shops closed. America's political elite attended the burial, and senators and cabinet members served as pallbearers. Bryan's former foes in the press now hailed his passion and integrity. "He tried to do the right thing as he saw it," the *New York Herald Tribune* observed in a typical editorial. "His passing will be a profound shock to millions who, however often he misled them, looked upon him as their prophet and counselor," a Philadelphia paper added.[29] This popular reaction was captured in a country music ballad, "Death of William Jennings Bryan," recorded only a week after the event; successive stanzas praised the Commoner for battling capitalists and helping workers, but one related directly to the Scopes trial: "He fought the evolutionists and infidel men, fools/ Who are trying to ruin the minds of children in our schools." Bryan clearly retained a large following in spite of (and perhaps due to) his role in the trial. Indeed, in a memorial tribute, William Bell Riley described the Tennessee trial as "Bryan's best and last battle."[30]

Many of those mourning their Peerless Leader's passing vowed to carry on his final crusade. On his way to the funeral, Bryan's brother Charles—a former governor of Nebraska and 1924 Democratic vice presidential nominee—talked of continuing his brother's work and forecast that "Congress eventually will be called upon to take a hand in the evolution controversy."[31] As Bryan lay in state, the governor of Mississippi declared that his state "will probably follow the lead of Tennessee and bar the teaching of evolution in the schools"—a prediction that came true as soon as his state's legislature next met. The public campaign for that legislation even featured speeches by Ben McKenzie telling the story of Bryan's last stand at Dayton.[32] During the fall of 1925, Texas Governor Miriam Ferguson—the South's first woman governor—directed her state's textbook commission to delete the theory of evolution from high school texts, a ban that for decades forced publishers to produce special edited versions of their biology textbooks for use in the Lone Star State. Throughout the country, dozens of fundamentalist leaders (including Riley, Straton, Norris, and Martin) rushed to don Bryan's fallen mantle, loosing a frenzy of uncoordinated antievolution activity. During the next two years, measures to restrict teaching evolution surfaced in more state legislatures than ever before.

A popular but short-lived legend began to develop (particularly in southern Appalachia) portraying Bryan as wholly triumphant at the

Scopes trial. Even *before* the Commoner's death, one correspondent wrote from Dayton at the conclusion of the trial, "For thousands in this section it would have come as no surprise if Mr. Bryan, having gloriously defeated the forces of unrighteousness, were to be visited by an angel of the Lord who would whisk the old gentleman off to Heaven in a chariot of fire." Less than a month after the verdict, an antievolution resolution adopted by the Clear Creek Springs encampment (a regional fundamentalist gathering in Kentucky) referred to the fight "so nobly and courageously led by the late William Jennings Bryan." Writing about Bryan's "conquest" at Dayton for the WFCA's journal, William Bell Riley asserted that the Great Commoner "not only won his cause in the judgment of the Judge; in the judgment of the jurors; in the judgment of the Tennessee populace attending; he won it in the judgment of an intelligent world."[33]

Country and western music captured this legend from the grass roots. The well-known Georgia balladeer Andrew Jenkins sang:

> There was a case not long ago in sunny Tennessee,
> The Bible then on trial there must vindicated be, . . .
> Oh, who will go and end this fight, oh, who will be the man?
> To face the learned and mighty foe, and for the Bible stand?

Bryan, the song continued, became the hero who "went to end the fight." Also in 1925, Columbia record company released a country folk song that declared:

> When the good folks had their trouble down in Dayton far away,
> Mr. Bryan went to help them and he worked both night and day.
> There he fought for what was righteous and the battle it was won,
> Then the Lord called him to heaven for his work on earth was done.

These and other compositions, collectively known as "Scopes songs," revealed a popular perception of the trial already departing sharply from the historical record. In this version of the Scopes legend, a heroic Bryan inevitably saved the schoolchildren of Tennessee from the damnable teaching of evolution. By September 1925, a popular-music magazine reported that the Columbia release "is selling exceptionally well, especially in the South and throughout the regions where the late Commoner was most active."[34] Three months after the trial, Mencken could only sneer about the man he so despised, "His place in the Tennessee hagiocracy is secure. If the village barber saved any of his hair, then it is curing gall-stones down there today."[35] The legend faded with time, but not before its inspiration led to the realization of Walter White's dream of a fundamentalist college in Dayton named in Bryan's honor. The Commoner and the Scopes experience had transformed Dayton from a religiously apathetic community into a center of faith.

At the time, in sharp contrast with later legends about the Scopes trial, no one saw the episode as a decisive triumph for the defense. "In examining the coverage of the trial in five geographically scattered newspapers and over a dozen national magazines," Ronald L. Numbers observed, "I discovered not a single declaration of victory by the opponents of antievolutionism, in the sense of their claiming that the crusade was at an end." Indeed, following Bryan's death, many of them feared precisely the opposite. A mid-August editorial in the *Nation*, a liberal journal of political opinion, referred to the antievolutionists' "success at Dayton" and predicted a "flood" of fundamentalist lawmaking across the land. In October, Mencken darkly warned, "The evil that men do lives after them. Bryan, in his malice, started something that it will not be easy to stop." Maynard Shipley's Science League took on the role of a Jeremiah by issuing a steady stream of dire prophecies about a pestilence of antievolutionism sweeping the nation. "The forces of obscurantism in the United States are in open revolt!" Shipley wrote, a full two years after the Scopes trial. "Centering their attacks for the moment on evolution, the keystone in the arch of our modern educational edifice, the armies of ignorance are being organized, literally by the millions, for a combined political assault on modern science."[36]

From the moment that he objected to courtroom prayer, a noisy segment of those opposed to antievolutionism blamed Darrow for the failure of the Scopes defense to stem the tide of fundamentalism—criticism that only increased after Bryan's death. Their reasoning reflected their own religious viewpoint. Some secular critics of the Tennessee law defended Darrow, but the Chicago agnostic was as much a pariah to religious modernists and mainline Christians as he was to fundamentalists. "Now that the chuckling and giggling over the heckling of Bryan by Darrow has subsided it is dawning upon the friends of evolution that science was rendered a wretched service by that exhibition," Walter Lippmann wrote for the *New York World.* "The truth is that when Mr. Darrow in his anxiety to humiliate and ridicule Mr. Bryan resorted to sneering and scoffing at the Bible he convinced millions who act on superficial impressions that Bryan is right in his assertion that the contest at Dayton was for and against the Christian religion." Speaking from the region most directly affected, the New Orleans *Times-Picayune* commented, "Mr. Darrow, with his sneering, 'I object to prayer!' and with his ill-natured and arrogant cross-examination of Bryan on the witness stand, has done more to stimulate 'anti-evolution' legislation in the United States than Mr. Bryan and his fellow literalists, left alone, could have hoped for."[37]

A varied array of Christians complained of Darrow's role in the case. "Some of us regret that in the unique Scopes trial, on which so much popular education depends, we should not have, for the trial, lawyers who understand the point of view of the ordinary, thoroughly educated Christian," a mainline church journal commented. A Congregational Church official offered his observations in a letter sent to the ACLU: "May I express the earnest opinion that not five percent of the ministers in this liberal denomination have any sympathy with Mr. Darrow's conduct of the case?" From Nashville, the Vanderbilt University humanist Edwin Mims complained, "When Clarence Darrow is put forth as the champion of the forces of enlightenment to fight the battle for scientific knowledge, one feels almost persuaded to become a Fundamentalist." The gravity of the issue was underscored for ACLU officials in September, when Raymond B. Fosdick—attorney for the Rockefellers and the brother of Harry Emerson Fosdick—curtly rejected an invitation to join their Scopes defense fund committee with the rebuke, "Largely through the choice of counsel a great opportunity was lost to place the case on a plane where the tremendous issue between tolerance and intolerance could be clearly seen." Roger Baldwin wrote back to Fosdick, "I guess we here all feel as you do about the handling of it. What we are trying to do now is to get the issue clear on the appeal."[38]

As Baldwin's letter to Fosdick suggests, ACLU officials in New York already had turned their attention to the appellate phase of the case. Once again they maneuvered to exclude Darrow from the defense team. The ACLU's executive committee hatched its plan in early August: ease out Darrow by urging more "priority" for Tennessee counsel. Associate Director Forrest Bailey, who oversaw the Scopes case for the ACLU, wrote to John Neal, "All of us feel that when the case goes to the Supreme Court of the State, there ought to be more of Neal and less of Darrow. This is the stand we are taking in response to the many urgent suggestions that Darrow ought to disappear from this point on." Hays dissented from this position, Bailey noted, "but I hope you see the other angle of the thing as we do." Apparently he did. Rappleyea soon reported to the ACLU about a meeting with Neal in which the two agreed that Darrow should not appear before the state supreme court. Otherwise, Rappleyea explained, "It will not be our cause on trial, but it will be a case of the State of Tennessee vs. Clarence Darrow, the man who spiritually and literally crucified Bryan on the cross examination."[39]

As chief counsel of record, Neal had the authority to remove Darrow from the case. To further encourage this step, Bailey wrote to several liberal religious leaders with close ties to the ACLU, asking them to write Neal concerning the matter. "We are constantly receiving criticism and protests concerning the Darrow personality and the harm it may do us," Bailey confided in these letters. "If your opinion is in accord with that which I have expressed, would you be willing to write a letter letting it appear that you do so purely on your own initiative, either to Mr. Darrow or to Dr. John R. Neal, expressing your personal convictions?" Within days, a pointed editorial appeared in *The New Republic*—whose editors enjoyed close ties to the ACLU leadership—criticizing the conduct of the Scopes defense and urging that Tennessee lawyers handle the appeal.[40]

It proved impossible, however, for the ACLU to secure either less of Darrow or more of Neal. The causes overlapped. Darrow soon learned of the ACLU's conspiracy against him and let union officials know that he knew about it. Bailey apologized with a letter dismissing the matter as a "misunderstanding" over the role of local counsel and lying, "I never at any time asked that you be invited to withdraw." Darrow shrewdly accepted Bailey at his word and even agreed with him about the need for "a leading Tennessee lawyer" to assist with the appellate argument, but added that Neal—Darrow's potential rival for control of the case—could not fill the role. "He is a fine man," Darrow wrote in strict confidence, "and could have been a good lawyer if he had given his time to it, but he has chosen to be a professor and is not equipped to take the leading part in arguing the case." Darrow suggested that the role go to either Robert S. Keebler, the Memphis attorney who had led the fight against the antievolution statute within the state bar association, or Frank Spurlock, the Chattanooga lawyer who had come to Darrow's defense in the contempt proceedings.[41] Bailey did not so easily give up on Neal either as a lawyer or as a means to assert control over the defense, and twice arranged for him to come to New York to discuss a diminished role for Darrow in the case. Neal turned back on both occasions before meeting with ACLU officials, apparently unwilling to confront Hays over the matter. Instead, he wasted his time filing two frivolous petitions in federal court seeking to restrain enforcement of the antievolution statute.

In early September, ACLU counsel Walter Nelles proposed to Bailey a different way to displace Darrow: Have the ACLU relinquish control of the case to a committee of prominent attorneys, and have it select the former U.S. Supreme Court justice and 1916 Republican presidential nominee Charles Evans Hughes to argue the appeal. Hughes soon signaled his willingness to join the cause with a thundering attack on fundamentalist lawmaking in his presidential address to the 1925 annual meeting of the American Bar Association. When Bailey shared Nelles's suggestion with Hays, however, the Scopes co-counsel exploded. "I think the effect of the publicity of the case and the manner of the trial gives us a better chance for reversal than if the matter had been handled otherwise," Hays wrote in an angry letter to Nelles. "For other lawyers to win it on appeal will take from Darrow and Malone the credit to which they are entitled. More than this, . . . I am not willing to have conservative lawyers and conservative organizations reap the benefit of work done by liberals or radicals." In a pointed reference to Hughes, Hays added, "I never yet have found any conservative lawyer who, at the beginning, wanted to undertake a case which *might* reflect discredit on him. When it turns out differently and there seems to be some publicity or honor to be had, then offers of assistance come from all over the country." Nelles backed down for the present, but asked Hays at least to consider alternative counsel should the case reach the U.S. Supreme Court.[42]

Darrow had other influential allies in addition to Hays, and their support (coupled with Neal's ineptitude) probably saved his place on the defense team. Several journalists, particularly those such as H. L. Mencken and Joseph Wood Krutch, who shared Darrow's hostility toward fundamentalists, maintained that the defense needed to expose Bryan and his "fool ideas" on religion, as Darrow called them at trial, even if it meant losing the case and alienating some mainline Christians. Furthermore, Watson Davis's Science Service arranged for all of the scientific expert witnesses to sign a letter endorsing Darrow's "ability, high purpose, integrity, moral sensitiveness, and idealism."[43]

The ACLU could hardly dismiss Darrow without antagonizing one faction of its supporters, and Neal seemed incapable of doing the job for them. Although Neal proudly boasted of his status as chief counsel, he persistently failed to communicate with co-counsel and missed the deadline for filing the bill of exceptions with the state supreme court. The missed filing precluded the defense from appealing any issues relating to the conduct of the trial—including the ruling on expert testimony; it could only challenge the validity of the antievolution statute. Even Bailey despaired and, in conjunction with Darrow, asked Scopes to bring Keebler or Spurlock on board as local counsel.[44] Scopes showed no interest in involving himself in the dispute, however, leaving the appeal adrift for months until Keebler, Spurlock, and other sympathetic Tennessee attorneys simply assumed responsibility for handling local administrative and procedural matters. "It was necessary to take control from Judge Neal," Darrow informed the ACLU office. "We never would have gotten the case to the Supreme Court unless we had taken steps to work without him." Amidst bitter infighting, the basic issue of control remained unresolved for the defense throughout the appeal process. As late as December, when one of the assisting local attorneys asked ACLU officials about who would argue the appeal, Bailey confessed uncertainty. "We ourselves have no interest in wishing to have Darrow, Malone and Hays continue," he answered. "My only point was that we should not be made to appear as having kicked these men out."[45]

Fundraising for the defense also languished. The ACLU had run up $5,400 in expenses for the Scopes case by year's end (mostly for expert witnesses), but it had raised only $3,800 in its special Tennessee Evolution Case Defense Fund, more than one-third of which came through

Malone. Although acting ACLU chair John Haynes Holmes assembled an impressive group of academics to serve on the fund's advisory committee, many of the "money people" (as Bailey called them) hesitated to join owing to Darrow's connection with the case.[46] Ultimately, it took a special appeal to members of the American Association for the Advancement of Science to wipe out the deficit, but that did not occur until 1926.

Internal conflict and confusion also hampered the state's effort on appeal. Primary responsibility for defending the law before the state supreme court fell on Tennessee's elected attorney general, Frank M. Thompson, but Governor Peay insisted on playing a part. Both officials, however, suffered from chronic illnesses that inhibited their ability to cope with the stressful case and would soon kill them. Peay spent the entire trial resting at a sanitarium in Battle Creek, Michigan, where he became seriously agitated over national coverage of the event. "While Governor Peay was still in Battle Creek," Nashville attorney K. T. McConnico later explained, "General Thompson conferred with me about my appearance as special counsel for the state in the Scopes case, which he said was needing study and attention more than he and his assistants could give it in view of the very onerous duties of his department and in view of his own physical condition."[47] Yet only the governor held authority to employ special counsel for the state, and Peay picked the Nashville attorney Ed T. Seay to handle the appeal. These two lawyers ended up working together on the matter, but at a high cost that the state balked at paying after Thompson and Peay died in office.

Further complications arose when William Jennings Bryan, Jr., and Samuel Untermyer asserted their interest in assisting the state. The elder Bryan had invited both of these attorneys into the case, but Peay and Thompson did not want outside counsel involved. "The people of this state thoroughly resented Darrow, Malone and the others from coming here to undo their statute and in deference to their feelings I suggested that it would be better for us to use local counsel," Peay wrote to Untermyer, who, ironically, had just joined the ACLU national committee but stood with Tennessee on the Scopes case.[48] Peay made an exception for the younger Bryan, largely out of sympathy, but the state later proved reluctant to reimburse his expenses.

With neither side pushing for early consideration of the matter, the appellate process dragged on for eighteen months, during which period the Scopes case continued to rankle Tennessee. At its annual meeting in November, the Tennessee Academy of Sciences went on record against the antievolution statute and soon thereafter filed a brief with the supreme court on Scopes's behalf. The Unitarian Laymen's League, a national association that included members from Tennessee, also submitted a brief in support of Scopes—the only other group to do so. Late in 1925, the Tennessee Christian Students Conference, a modernist association of collegians from throughout the state, adopted a resolution condemning the antievolution statute as harmful to both education and religion.[49] On the other side, conservative religious and patriotic organizations besieged Governor Peay with letters and petitions urging him to stand firm. The Ku Klux Klan took up the cause with vigor, and the defrocked Klan official Edward Young Clarke formed a short-lived rival group in the Southeast called the Supreme Kingdom, whose primary purpose was carrying on Bryan's crusade against teaching evolution. Regional opinion had so solidified that when Mississippi passed its antievolution statute early in 1926 the ACLU could not entice a local teacher or taxpayer to challenge it, despite offers similar to those that recruited Scopes.[50]

Tennesseans caught in the middle felt increasingly frustrated. George Fort Milton, for example, took his family on an extended trip to the West Coast. "We hope to wipe out the last trace of the Dayton trial, for it was a trial not only of Scopes, and of the state, but of the fortitude and self-possession of all of us," he wrote to Peay, adding a plea "to keep down any attempts of your friends to inject fundamentalism into state politics as a political issue." Yet 1926 was an election year in Tennessee, with all state offices on the ballot—including every seat on the supreme court. "The greatest problem has been in keeping the case out of politics," one of its assisting local attorneys reported to the ACLU. "Both the Democratic candidates for governor are loudly proclaiming their respective defenses of the faith and Peay's only opponent is doing all within his power to out Herod Herod." Running on his record as both a progressive reformer and defender of religion, and despite declining health, Peay handily won an unprecedented third term as Tennessee governor, only to die a few months later. In an apparent attempt to keep the Scopes controversy out of their campaigns, the state supreme court justices delayed their decision in the case until after the election. Scopes privately interpreted this as a sign that they planned to overturn the statute. His lawyers shared his optimistic outlook.[51]

Defense counselors based their optimism on the strength of their written and oral arguments to the high court. Hays and Keebler took the lead in drafting the defense's appellate brief, signed by nine attorneys—including Darrow, Malone, and Neal. After reviewing the book-length document, Bailey expressed his concern to Hays that the names of "six 'foreigners' and only three 'natives' " appeared as signatories, but conceded to Keebler, "The brief is an excellent one and ought to do the trick." In reply, both attorneys smugly predicted victory, though Hays had admitted earlier to Nelles that "perhaps I have become over-convinced by the brief I have written."[52]

At least to themselves and their crowd, the defense's argument seemed compelling, even though it added nothing of consequence to that made at Dayton. Once again, the defense stressed that the anti-evolution statute unreasonably restrained the individual liberty of teachers and students by establishing a preference in public education for a particular religious belief over the conclusions of modern scientific thought. Since the argument had not changed, defense hopes clearly rested on an assumption that a more sophisticated audience in Nashville would judge it. Speculating about the impact of this brief on opposing lawyers, Neal gloated, "It is quite evident that they have been presented with a much tougher problem than they expected. Victory doesn't now seem quite so impossible or visionary."[53]

If the defense brief intimidated Tennessee's lawyers, their 400-page reply brief did not show it. Writing for the state, Seay and McConnico countered the defense's plea for academic freedom with an unabashed appeal to majoritarianism that would have made the Commoner blush. "The public schools are created by the legislature," the brief began in Bryanesque fashion, "and the courts can in no manner control, limit or proscribe the legislature in the exercise of power" over them. It did not end there. "The fact that a group of self-styled 'intellectuals' who call themselves 'scientists' believe that a certain theory or thing is true does not to any degree prevent the state legislature . . . from forbidding the teaching or practicing of such a thing or theory which the legislature may conclude to be inimical . . . to the general public welfare." Bryan crusaded only against teaching evolution in public education and maintained throughout that evolutionists could start their own schools; the state's brief recognized no such limits on majority rule. "'Scientific' *superficialists* and *intolerants*," it emphasized, "under a *perhaps* soiled or even red banner of 'academic freedom' [cannot] foreclose the *police power* of the State's constitutionally chosen and elected representatives as to what is required for the public welfare." Relying on a recent U.S. Supreme Court decision upholding a compulsory school vaccination program, the brief asserted, "What the public believes is for the common welfare must be accepted as tending to promote the common welfare, whether it does in fact or not."[54] In this brief, the state sounded more like Billy Sunday than Bill Bryan.

Briefs submitted by the Tennessee Academy of Sciences and the Unitarian Laymen's League attempted to answer the state's arguments, as did a supplementary memorandum filed by the defense. "Suffice it to say here that . . . their theory would absolutely nullify constitutional government and inaugurate the dictatorship of the majority," the Academy's brief noted. "The cases relied on by the State involved reasonable regulations made by the legislative body in relation to public work—not unreasonable, arbitrary, and capricious regulations." All three additional appellate documents attempted to show the unreasonableness of the antievolution regulation, the Academy stressing scientific arguments for teaching evolution and the Unitarians attacking religious ones against it. "Innumerable numbers of our greatest Christian scientists, philosophers, educators and ministers firmly believe in the truth of the origin of man as taught by evolution," the Academy's brief concluded. "The State has no right . . . to stifle by legislation the influence of such men."[55]

The Tennessee Supreme Court set aside two days at the end of May for oral arguments—much more time than it customarily gave to cases. Scopes's trial team reassembled for the event, augmented by Keebler but absent Malone; Seay and McConnico stood alone for the state. "Newspaper men from many parts of the country are assembling in Nashville, and special leased wires have been arranged by the various news agencies to lead from the courtroom itself," one local paper reported. Spectators again filled the chamber. "Every door and window was blocked by scores who, unable to gain entrance, contented themselves by standing on chairs and tables," another paper noted. "Many others are turned away." Darrow promised to keep on his big-city suit coat and vest this time, but assured reporters that he still wore the small-town suspenders underneath, joking that he "would doubtless be quite lost without them." The two most responsible for bringing the case, George Rappleyea and John Butler, claimed front-row seats. Scopes himself refused to attend, curtly telling the press that he was "not interested in the outcome and wanted to forget the entire episode." Promoters and proselytizers had harassed him for nearly two years; he wanted his privacy back.[56]

Fulfilling his now titular role as chief counsel for Scopes, Neal rose first to introduce the case. It soon became clear, however, that the defense never resolved who should deliver its oral argument. As it turned out, everyone would. An attorney for the Unitarian Laymen's League opened the argument with a rambling denial "that the teaching of the evolution theory is likely to cause our youth to lose their faith in God." Hays followed with a thumping legal plea grounded squarely on the due process clause of the Fourteenth Amendment, which he claimed prohibited any state from enforcing unreasonable laws. Tennessee's "absurd" antievolution statute violated this standard as much as a law against teaching Copernican astronomy would, Hays asserted. "The theory of our constitution is that in the competition of ideas, truth will prevail," he concluded. "We plead for freedom of education, for the liberty to teach, and the liberty to learn, for in this small statute lies the seed of a doctrine which in generation may reach out and stifle education." Counsel for the Tennessee Academy of Sciences finished the initial presentation for the defense by warning of dire consequence for science and medical education if the law remained in effect.[57]

Seay opened the state's rebuttal with a broadside. "Our adversaries tell you it is a controversy between modernists and fundamentalists. But I tell your honors there is something more," he warned. "If you permit the teaching that law of life is the law of the jungle, you have laid the foundation by which man can be brought to accept the doctrine of communism and to the point where he believes it right to advocate murder." Seay identified Darrow as a case in point: a defender of Communists and murderers as well as evolutionists. "The Tennessee legislature passed this law to stamp out worse things" than merely teaching evolution, he baldly asserted. Yet the statute itself did not cross the line into promoting religion, he assured the court, because "there is no authority for any teacher to teach the theory of divine

Arthur Garfield Hays at the time of the Scopes litigation. (Courtesy of Bryan College Archives)

creation in the public schools." Seay thus gave the statute a clear secular purpose rather than a solely religious one. Seay reinforced his remarks by reading from a written statement submitted to the court by William Jennings Bryan, Jr., in which he described the law as the "deliberate, thoughtful enactment of a sovereign people, which was designed to protect their children in their own public schools in their beliefs in the divine origin of man, which in turn measures their responsibilities to God and their fellow man." Borrowing from the rhetorical arsenal of the defense, the younger Bryan described the issue as one of "freedom" but defined that concept in a way his father would have recognized: the "freedom" of the majority to "protect the children of the state in the public schools in their common belief."[58]

The court adapted its customary procedure to accommodate the unusual case by permitting Keebler and co-counsel from the Academy of Sciences to respond to Seay before hearing the scheduled closing arguments. Resurrecting accusations leveled against the ACLU during the Red Scare, Seay attributed the ACLU's interest in the Scopes case to a pro-Communist agenda. Keebler countered that such interest arose solely because the statute denied academic freedom. The law did not serve the broad secular purpose of protecting American values, he added, but merely promoted "that peculiar dogmatism of the Christian church" known as fundamentalism.[59] After court recessed for the day, the Academy co-counsel resumed this line of defense in the morning. He denounced Seay's personal attacks on Darrow and evolutionists. "We all know that thousands of people in this very state, who are devoutly Christian, believe at the same time in the divine origin of man and in his development from a lower form of life," the Nashville attorney added. "There is obviously no conflict between the doctrine of evolution, as applied to man, and the doctrine of man's divine origin." After hearing these lawyers, one reporter observed, "Whereas in Dayton the Scopes defense found little sympathy for their arguments among the town's citizens and less still from the members of the bar, here the people are indifferent while the local attorneys express convictions that the law is invalid."[60]

Any public indifference regarding the law's validity did not diminish popular interest in the case, however. In anticipation of hearing Darrow's closing argument, a record audience assembled for the second day, including "many women of prominence in the social life of the city," as one reporter observed. McConnico, a noted local orator, arrived early for his clash with Darrow and brought with him a large selection of his opponent's published writings. When Darrow arrived and saw his "Argument in the Defense of Communists" laying on top of the pile, he joked with McConnico, "You won't have much time for anything else."[61] The scene was set for the dramatic close.

Hoping to establish guilt by association, McConnico resumed the state's broad assault against the defense. He stressed Darrow's agnosticism and the ACLU's radicalism. When seeking aid in challenging the statute, McConnico charged, "Dr. Neal knew where to get comfort." The ACLU's promotion of the trial and Darrow's treatment of Bryan received severe criticism. When the state's counsel began criticizing local attorneys for assisting the defense, the court finally directed him to "confine himself to the case before the court." By then, however, his allotted time was running out. He closed with two points. In passing the law, McConnico asserted, the legislature simply sought to "preserve the Bible for all sects" rather than to favor fundamentalism. Ninety-five percent of Tennesseans believed in life after death, he estimated, and teaching evolution undermined this vital faith. Furthermore, McConnico added, the statute did not establish religion but merely provided that "since the Bible can't be taught [in public schools], we wouldn't let this thing called anti-Bible be taught." The state's attorney asked the court, "Would not Tennessee be committing a tragedy in civil government if it did not intervene to prevent the teaching in her public schools of a dogma conceded to destroy the minds of the people, whether it is right or wrong?" To cheers from the gallery, he urged the justices "to resist the 'sinister and unclean' efforts being made to teach 'this animal dogma' in the public schools of the state."[62]

Ignoring the personal attacks against him, Darrow delivered his prepared closing. It was less a legal argument than a plea for freedom that reflected a thoroughly modern view of science and religion. Seay and McConnico had referred to the Bible as truth and science as opinion; Darrow now reversed the designation. He portrayed religion as a personal matter "that ought to be the affair of the individual" and science as a public activity that "is the cause of progress . . . and everything that makes civilization today." In accord with his viewpoint, he asserted, "The schools of this state were not established to teach religion. They were established to teach science." Darrow assumed that the Tennessee and U.S. constitutions, through their protection for public speech and from religious establishment, shielded science from religion in the common schools. "The future of America's public school system and the complete education of her children can be safeguarded only by wiping this law off the statute books," he declared. When asked by one justice if the state could bar all teaching about origins (whether religious or scientific), as Seay and McConnico proposed, Darrow replied, "You would have to first amend your school act which prescribes that biology must be taught because that is all biology is." Comparing Scopes's conviction to the execution of Socrates, Darrow concluded his hour-long argument to great applause as he declared, "We are once more fighting the old question, which after all is nothing but a question of the intellectual freedom of man."[63]

Despite rhetorical flourishes on the final day, the appellate hearing failed to stir passions. "Argument before the Supreme Court was in a far different atmosphere from that of the trial," Hays observed. The *Chattanooga Times* called it "a flop as a news story as compared with the trial." At the end, an Associated Press correspondent observed, "Those who had been drawn to the courtroom in hopes of hearing a verbal clash between the distinguished counsel were disappointed." The formal nature of the proceedings certainly stifled emotions. "Everything was calm, dignified and quiet," Hays recalled. "There was no rising to the feet to interpose objections, no bickering between members of the counsel and no religious or anti-religious atmosphere," the *Chattanooga Times* added.[64] Bryan's absence reduced the drama and Darrow's style best fit a trial. Finally, the hearing did not end with a climactic verdict. The high court simply took the arguments under advisement and, in this case, waited seven months before issuing its opinion. In the meantime, both sides forecast victory.

Belying its public predictions, the ACLU privately planned for the case to continue. Those plans did not include Darrow. "Now that the Scopes case has gone up to the Tennessee Supreme Court, it is time to consider policies in connection with further appeals," Bailey wrote to Darrow two days after oral arguments. "We feel that whoever argues the case before the United States Supreme Court should be utterly beyond the reach of prejudice of certain members of that august body, and we seriously doubt whether you, Mr. Malone or Mr. Hays, for example, would meet this requirement." Flatly refusing to step aside,

Darrow shot back, "Any possible prejudice that might exist to Mr. Hays or me, would be very much stronger against your organization." ACLU leaders stood firm. This appeal meant too much to them: a chance for the ACLU's first major victory and a crack at respectability with Charles Evans Hughes on board. They appealed to Scopes. "I want to say confidentially now that if Mr. Darrow and Mr. Hays insist upon staying in the case and arguing it before the U.S. Supreme Court, the Civil Liberties Union will probably have to withdraw," Roger Baldwin informed Scopes. "You are the defendant. You have the right to employ whom you choose to take up your case. . . . You may be put into the position of making the choice." ACLU counsel Wilcott H. Pitkin privately explained that Darrow's continued participation "would probably influence badly one or more judges of the Supreme Court [and] would make it impossible . . . to procure the names of reputable church-going lawyers on the brief."[65]

Although he later said that he would have stuck by Darrow, Scopes never had to make the choice.[66] In a clever maneuver, the Tennessee Supreme Court managed to end the embarrassing case without overturning the locally popular law. The antievolution statute only applied to public employees acting in their official capacity, and therefore did not infringe on individual liberty, the court ruled. Scopes "had no right or privilege to serve the state except upon such terms as the state prescribed." Furthermore, the court added, the law "*requires* the teaching of nothing," and therefore "we are not able to see how the prohibition . . . gives preference to any religious establishment." Accordingly, three justices—a bare majority—held the statute constitutional; only two of them, however, interpreted it to bar all public school instruction in human evolution. One justice concocted an exemption for theistic evolution out of the statute's confusing language. Another called the entire law "invalid for uncertainty of meaning." The court's fifth member died before the ruling, and his successor took no part in the decision.[67]

After upholding the statute, however, the court overturned Scopes's conviction on the grounds that the trial judge, rather than the jury, fixed the amount of the fine. Yet Raulston had simply imposed the minimum fine after offering the jury an opportunity to set a higher one. Both parties had accepted this procedure at trial and neither raised it on appeal. That should have settled the issue, but the court used it to reverse the conviction, then urged the attorney general to dismiss the prosecution. "We see nothing to be gained by prolonging the life of this bizarre case," the court wrote. "On the contrary, we think the peace and dignity of the state . . . would be better conserved by the entry of a *nolle prosequi* herein."[68] Without comment, Tennessee's new attorney general complied a day later—which left no conviction for the defense to appeal.

The most widely publicized misdemeanor case in American history had finally come to an end, with neither side claiming victory. The defense immediately cried foul. Malone denounced the ruling as "a subterfuge on the part of the State of Tennessee to prevent the legality of the law under which Scopes was convicted being tested" by the U.S. Supreme Court. "The whole matter is left in an unsettled condition," Darrow complained. "It will probably require another case to clear up the matter." Scopes simply dismissed the decision as "a disappointment," and no other Tennessee teacher ever again stepped forward to challenge the statute.[69] Supporters of the statute, however, could scarcely hail a ruling that all but directed prosecutors not to enforce the law. "Some public officers," the *Nashville Banner* reported, "were of the opinion that the state was 'fed up' over the Scopes case, and that no circuit attorney-generals would care to reopen the question by bringing indictments."[70] None ever did. The antievolution statute became precisely what Governor Peay predicted when he signed it into law—a largely symbolic act.

Partly as a result of the Scopes trial, the law came to symbolize different things to different people; it became a symbol of pride and regional identity for some Southerners. An Alabaman, for example, wrote to Peay, "The Great Commoner fell at this post maintaining that Tenn. had sense enough to run her own affairs without Yankees from the North to meddle in them."[71] Shortly after the Tennessee Supreme Court ruling, Maynard Shipley wrote of "the threatened South," where antievolutionism still constituted "a serious menace."[72] By popular referendum, the people of Arkansas enacted the region's third antievolution statute in 1928. Louisiana soon thereafter joined Texas in barring any mention of evolution from state-approved textbooks. School boards throughout the South imposed local limits on teaching evolution. Subsequent studies suggested that such restrictions enjoyed widespread support among the various groups that comprised southern society.[73]

At the same time, the tendency of northern evolutionists to blame Southerners for the Scopes trial may have weakened antievolutionism in the North. For example, patrons at New York's famed *Ziegfeld Follies* cheered Malone's returning declaration, "Although we went South, we insisted upon retaining our Northern ideas." In a typical posttrial comment, one northern journalist linked "the inquisition in Tennessee" to "the South [as] a cultural wilderness." H. L. Mencken continued to tar and feather the South over the Scopes trial for years, leading Edwin Mims to compare the Baltimore journalist with the Civil War general William Tecumseh Sherman for his treatment of the region. Northerners tended to laugh along with Mencken's satire and after the Scopes trial displayed little interest in adopting antievolution laws. When one Rhode Island legislator introduced such a proposal in 1927, his bemused colleagues referred it to the Committee on Fish and Game. Even though the antievolution crusade began as a legitimate national movement, with such urban Northerners as Riley and Straton in the lead, it became for the most part a regional phenomena after the Scopes trial. Reporting on the failure of all efforts to repeal the myriad restrictions against teaching evolution in the South, the ACLU attributed it to "stubborn Southern hostility against Northern conceptions of science and faith."[74]

During the years immediately following the Scopes trial, partisans on both sides battled over its legacy. Darrow, Hays, and Malone lampooned the prosecution in books, articles, and lectures. They made much of Bryan's alleged concession on the witness stand that the biblical days of creation symbolized long periods of time, Hays claiming that "even for Mr. Bryan our case might have been proved" and Darrow crowing that "Bryan had contradicted his own faith."[75] Popular science writers such as Henry Fairfield Osborn assailed the prosecution's ideas of biology and science education. Several of the defense expert witnesses wrote semipopular books or articles expanding on their trial affidavits. Such accounts leave the distinct impression that Scopes *won* the case in all but the verdict, which "hillbilly" jurors withheld.

Fundamentalists countered in publications and presentations of their own. Riley wrote a long article in the WFCA's journal refuting defense arguments from the trial and defending Bryan's testimony. For example, he explained to his followers, "Imagine a self-respecting attorney . . . by every conceivable twist and turn trying to get Mr. Bryan to say that 'God created the world six thousand years ago,' when the plain statement of Genesis . . . leaves latitude for millions and even billions of years."[76] Darrow's villainy at trial became the subject of countless fundamentalist sermons. In 1927, the popular antievolution science lecturer Arthur I. Brown answered Osborn's most famous trial-related article, "The Earth Speaks to Bryan," with his booklet, "Science Speaks to Osborn." Other opponents of teaching evolution dissected the defense's scientific affidavits, apparently on the assumption that these documents represented the case for evolution in the popular mind. These fundamentalists made it sound as if the evolutionists suffered a total rout at Dayton.

So long as the antievolution crusade raged on and partisans battled over the trial's interpretation, historians hesitated to assess the significance of events at Dayton. The trial slipped into the tail-end of Charles A. Beard's monumental co-authored survey, *The Rise of American Civilization*, the final volume of which appeared in 1928. Beard at that time stood atop America's left-leaning history establishment and rarely passed up a chance to stamp a dialectical interpretation on events. This work identified the trial as a "spectacular battle" in an ongoing "war" between mostly rural fundamentalists and urban modernists but did not present the battle as decisive or the war as resolved. "Among the freethinkers of two continents," it observed, "the Tennessee case aroused amusement at the expense of the American hinterland, but undisturbed by scorn from such quarters, the Fundamentalists announced that they intended to carry on their campaign."[77] The historian Preston William Slosson included two paragraphs on the trial in his pioneering history of the twenties, *The Great Crusade and After*, yet failed to comment on its significance. "The trial resolved itself into a verbal duel between Darrow the agnostic and Bryan the Fundamentalist," he wrote, but Bryan died "a martyr in defense of the faith" and "no one, not even Clarence Darrow, was quite big enough to inherit the mantle of Tom Paine or Bob Ingersoll as a popular American champion of anticlericalism." The account ended with an unanswered question, "What was the actual state of religious faith among the American public?" Although the Scopes trial secured a bit part in American history, it still lacked a decisive role.[78]

As Slosson's unanswered question suggests, historians at the end of the twenties did not perceive any slowing in the pace of fundamentalist political activism. When that trend became apparent over the next decade, some historians began attributing it to the movement's alleged defeat at Dayton. As late as 1930, the ACLU and the Science League continued to issue grim bulletins about antievolution activity, however. At most, intellectuals saw the trial as a personal humiliation for Bryan. Both Hays, in 1928, and Darrow, in 1932, took this approach in their autobiographies, and Mencken proclaimed it in his writings. In 1929, two debunking biographies of Bryan presented such a view as history, one asserting that Darrow made "hash" of the Commoner.[79]

None of the works cited suggest that the debunking of Bryan would slow the antievolution crusade, because liberal commentators of the day typically viewed fundamentalists as servile rubes immune to shame. In his highly critical 1931 *History of Fundamentalism*, for example, Stewart G. Cole ridiculed Bryan's role in the Scopes trial, yet wrote that it *quickened* the pace of antievolution activity, especially after Riley, Straton, and others began competing to carry on the Commoner's work. On balance, early historical reflection on the Scopes trial presented the episode as a sign of the times rather than as a decisive turning point.[80] At the decade's end, the legacy of the Scopes trial remained up for grabs.

CHAPTER NINE

RETELLING THE TALE

THE MODERN SCOPES legend emerged during a thirty-year period bracketed by the appearance of two enormously popular creative works. The process began in 1931, when *Harper's* magazine editor Frederick Lewis Allen published his surprise best-seller, *Only Yesterday: An Informal History of the Nineteen-Twenties,* and culminated in 1960 with the release of *Inherit the Wind*, a popular motion picture based on a long-running Broadway play by Jerome Lawrence and Robert E. Lee. Far more than anything that actually happened in Dayton, these two works shaped how later generations would come to think of the Scopes trial.

In writing *Only Yesterday*, Allen never specifically intended to shape public perceptions of the trial; as the country sank into the Great Depression, he simply sought to relate the happier days of the Roaring Twenties in a lively, journalistic fashion. Without any formal training in historical methods but with a reporter's knack for chronicling events, Allen drew up a calendar for the decade and used old almanacs and periodicals to fill in top news stories for each month. He then transformed his outline into a fast-paced narrative. As the major news story of mid-1925, the Scopes trial became the feature event of a middle chapter in Allen's book.

Allen presented the trial in cartoonlike simplicity. The growing "prestige of science" sapped the "spiritual dynamic" from modern America, he asserted. Fundamentalists, in reaction, clung to "the letter of the Bible and refused to accept any teaching, even of science, which seemed to conflict with it." Modernists, in contrast, "tried to reconcile their beliefs with scientific thought; to throw overboard what was out of date." Skeptics, "nourished on outlines of science," abandoned religion. "All through the decade the three-sided conflict reverberated. It reached its climax in the Scopes case in the summer of 1925," Allen wrote. "In the eyes of the public, the trial was a battle between Fundamentalists on the one hand and twentieth-century skepticism (assisted by Modernism) on the other." Subplots, he suggested, included "rural piety" versus urban sophistication and the South against the North. The defense's fight for individual liberty and the prosecution's appeal to majoritarianism disappeared from Allen's version of events, as did the ACLU and the WCFA. His account pit Darrow against Bryan in a bitter, farcical encounter set amid a media frenzy in the circuslike atmosphere of Dayton boosterism. "The climax—both of bitterness and of farce—came on the afternoon of July 20th," when Bryan "affirmed his belief" in various Old Testament miracles under Darrow's withering interrogation. "The sort of religious faith which he represented could not take the witness stand and face reason as a prosecutor," Allen concluded. "Theoretically, Fundamentalism had won, for the law stood. Yet really Fundamentalism had lost . . . and the slow drift away from Fundamentalist certainty continued."[1]

In *Only Yesterday*, Allen reduced fundamentalism to antievolutionism and antievolutionism to Bryan. Both reductions grossly oversimplified matters and forced Allen to reconstruct the story. For example, he wrote that under Darrow's questioning, "Bryan affirmed his belief that the world was created in 4004 B.C.," whereas Darrow actually wrung out a concession that the Genesis days of creation represented long periods of time, leading to Darrow's triumphant claim that "Bryan had contradicted his own faith." Also, Allen never mentioned Bryan's forced admission on the stand of his ignorance about science, which earlier commentators viewed as so important in debunking antievolutionism. Allen noted only the Commoner's blind faith in the Bible. Yet equating Bryan with fundamentalism enabled Allen to become the first published commentator to transform Bryan's personal humiliation at

Dayton into a decisive defeat for fundamentalism generally. Of course, he could not cite much hard evidence to support his claim that Americans were losing their religion. Indeed, he conceded, "If religion lost ground during the Post-war Decade, the best available church statistics gave no sign of the fact." He dismissed such statistics as superficial, however. "In the congregations," he maintained, "there was an undeniable weakening of loyalty to the church and an undeniable vagueness as to what it had to offer them."[2] This had been true in his own life, but to extrapolate it to all Americans—and to suggest that the Scopes trial contributed to the process—was sheer speculation.

Allen never claimed to offer more than he delivered. "This book is an attempt to tell, and in some measure to interpret, the story of what in the future may be considered a distinct era in American history," he wrote in the preface. "One who writes at such close range, while recollections are still fresh, has a special opportunity to reveal the fads and fashions and the follies of the time [and] . . . leave to subsequent historians certain events . . . the effect of which . . . may not be fully measurable for a long time."[3] This approach struck a responsive chord in the thirties. Allen geared his book for a popular audience and cautiously hoped for modest success, but nostalgia for the twenties propelled sales beyond his wildest dreams. It quickly became a best-seller and ultimately sold over a million copies—more than any other nonfiction book of the decade. More remarkably, it influenced historians and remained widely used as a college history text for more than half a century. "No one has done more to shape the conception of the American 1920s than Frederick Lewis Allen," the historian Roderick Nash later observed about Allen's book. It "has been the font at which most subsequent writers about the decade initially drank." Owing to Allen's method of "seizing on the decade's most glamorous aspects and generalizing from a few headlines," Nash added, "the book's most enduring bequest to later historians has been the idea that older American values, traditions, and ideals meant little or nothing to the 1920s."[4]

By portraying the Scopes trial as a decisive defeat for old-time religion, Allen fit the episode neatly into his general conception of the twenties as a time when America repudiated its Victorian traditions. As a result, readers accepting Allen's interpretation saw the trial as a step in the triumph of reason over revelation and science over superstition in modern America. Darrow might welcome such a verdict, but it did not particularly serve the interests of the ACLU, which had instigated the trial as a means to fight for freedom rather than against religion. In fact, Allen's presentation of fundamentalism as a vanquished foe frustrated ongoing ACLU efforts to portray it as a persistent threat to individual liberty. Moreover, it may have encouraged evolutionists to let down their guard. The Harvard biologist Ernst Mayr immigrated into the United States in the year that Allen's book appeared. "Looking back that far, my impression is that I thought that this trial was the end of the fundamentalist attacks on evolution," he later wrote. "I believe my interpretation was widely shared by American evolutionists. As a result not much time and effort was spent by evolutionists in America to prove the fact of evolution and to refute the claims of the fundamentalists."[5]

In addition to attributing a decisive outcome to the trial, *Only Yesterday* perpetuated various misconceptions about events at Dayton. For example, not only did Allen pass along an altered version of Bryan's trial testimony, he sharpened the entire episode. Darrow's drawn-out questions "about Jonah and the whale, Joshua and the sun," and the like, now appeared in rapid-fire succession, without any indication of Bryan's various answers, and the dramatic call for Bryan to take the stand occurred "on the spur of the moment," according to Allen, instead of as a carefully planned maneuver. He seriously confused the trial's origins. In his account, the idea for the lawsuit came from Rappleyea and Scopes, not the ACLU; Scopes then intentionally broke the law and "was arrested" while Rappleyea "secured for Scopes the legal assistance of Clarence Darrow" and others. Allen surely did not mean to distort the story—he simply relied on inaccurate news accounts and his own faulty memory. Through his book, this version of events passed into the Scopes legend.[6]

Later writers adopted Allen's verdict and accepted his depiction of events. Gaius Glen Atkins relied heavily on *Only Yesterday* in writing his semipopular 1932 account, *Religion in Our Times*. The journalist Mark Sullivan did the same for his 1935 best-seller, *Our Times: The United States, 1900–1925*. Both books presented the trial as the decisive event in the history of American fundamentalism. For Atkins, it "marked the furthermost advance of the movement." Sullivan called it the "explosive climax" of the fundamentalist controversy. Both portrayed Darrow's interrogation of Bryan as the turning point when, as

Atkins put it, religion was "made to look ridiculous." Science and critical thought triumphed, he concluded: "The Scopes trial marked the end of the age of *Amen* and the beginning of the age of *Oh Yeah!*"[7] William W. Sweet revised his widely used collegiate religious studies text, *The Story of Religion in America*, to reflect the new view. The 1930 edition of his book depicted the trial as a media event that did not reach the "broad issues" raised by fundamentalism. According to the 1939 edition, however, the trial resolved these issues in the public mind. "Bryan said evolution . . . made God unnecessary, denied the Bible and destroyed all belief in the supernatural," Sweet wrote. "Darrow attempted to make Bryan look ridiculous and submitted him to a mocking examination. It was Fundamentalism's last stand." The Scopes trial became a watershed event. Secular commentators generally concurred with the novelist Irving Stone's 1941 analysis that the Darrow–Bryan clash "dealt a deathblow to Fundamentalism."[8]

Although Allen unintentionally misinterpreted events leading up to and including the trial itself, ongoing developments led later commentators to follow him. By the 1930s, fundamentalist political activity had decreased to such an extent that outside observers thought the movement had died. The Scopes trial offers a convenient explanation for this development, but the timing doesn't quite fit. Riley, Straton, and other fundamentalist leaders initially perceived the trial as a victory for their side; none seemed despondent about it at the time. Furthermore, antievolution activism increased noticeably for several years following the verdict, with additional states imposing restrictions. Fundamentalist church membership continued to grow during the twenties and on into the future. While it is true that open warfare between fundamentalists and modernists quieted down during the late 1920s, and that the political crusade to outlaw teaching evolution ended by 1930, at most the Scopes trial contributed only indirectly to any apparent decline of fundamentalism.

Each side went to Dayton confident that a full airing of the issues would aid its cause. "I am expecting a tremendous reaction as the result of the information which will go out from Dayton," Bryan wrote shortly before the trial. The defense made similar pretrial predictions about its prospects, such as Scopes's observation, "There is no doubt in my mind that through this open, frank discussion, a better understanding will result." Each side left Dayton confident that it had achieved its objective.[9] On this issue, discussion did not resolve disagreement; each side so deeply believed in its position that further information simply increased its vehemence.

By focusing attention on the topic of teaching evolution, the Scopes trial encouraged both sides—with the result that, by the end of the decade, most states or localities where fundamentalists held political power had imposed antievolution restrictions by law, administrative ruling, or school board resolution. This included most of the South and some of the West. In the North, however, efforts to outlaw teaching evolution met with stiff resistance and humiliating defeat. In 1927 alone, antievolution bills lost in over half a dozen northern states. The most stunning setback occurred in Riley's home state of Minnesota where, despite an all-out blitz by fundamentalists, the bill lost by an eight-to-one margin in the state legislature. "This dismal failure was a crushing blow," Riley biographer William Vance Trollinger, Jr., wrote, and it "signalled the end of William Bell Riley's efforts to secure antievolution legislation."[10] The campaigning ended, however, only after it became obvious that each side had reached the geographical limit of its influence. All of the commentators who later pronounced that fundamentalism had died in Dayton (such as Allen, Atkins, and Sullivan) came from the North, where the trial had set back antievolutionism. Southerners saw it differently. During the thirties, for example, the North Carolina sociologist Howard W. Odum could still report about his region that "upon all questions, political, financial, educational, scientific, and technical, the judgement of religion and scripture was likely to be invoked."[11] Furthermore, once in place, no southern antievolution restriction was repealed for over forty years.

Alternative reasons existed for the decline in antievolution activity at the time. By the thirties, fundamentalists had less reason for concern about teaching evolution than before the Scopes trial. Not only had many states and school districts limited such instruction, but their restrictions influenced the content of high school biology textbooks everywhere. To serve the southern market and in response to heightened sensitivity about the topic, national textbook writers became increasingly less dogmatic in their presentation of Darwinism. This process began even before the trial. Worried about sales of its biology text, for example, one major publisher sought an endorsement from Bryan by offering to present evolution as a "theory" rather than "dogma." Bryan welcomed the suggestion, but responded, "It would take a great deal in the way of elimination and addition to make it clear that evolution is presented only as a hypothesis."[12] After the Scopes trial, many biology textbooks underwent such revision.

The evolution of George W. Hunter's *Civic Biology* exemplified the process. The Tennessee Textbook Commission dropped the book from its approved list shortly after Scopes's indictment for using it. A year later, the book's publisher deleted a six-page section on evolution from copies of the text sold in some southern states, and Hunter began work on revising the entire book. He cut out the section title, The Doctrine of Evolution, and deleted charts illustrating the evolution of species. A revised passage about the "development of man" looked back only to "races of man who were much lower in their civilization than the present inhabitants" rather than to subhuman species, and included the biblically orthodox addition, "Man is the only creature that has moral and religious instincts." A paragraph on "natural selection" remained, but with every sentence qualified as something that Darwin "suggested," "believed," or "said." Hunter no longer hailed Darwin as "the grand old man of biology," and the phrase about Darwin, "his wonderful discovery of the doctrine of evolution," became "his interpretation of the way in which all life changes." Indeed, the inflammatory word *evolution* disappeared altogether from the post-Scopes version of the text, and equivocation replaced certainty wherever evolutionary concepts remained. Other schoolbook writers followed Hunter's example.[13] By the time Hunter finished, antievolutionists had little grounds for complaint, though they scarcely would admit it.

Fundamentalists continued to complain about Darwinism, of course, even if they stopped crusading against teaching evolution. Moreover, fundamentalism did not die. To the contrary, it attracted an ever-increasing number of adherents nourished on a steady diet of antievolution books, articles, and tracts published by conservative Christian presses. Riley continued to churn out antievolution pamphlets long after he gave up crusading for antievolution legislation. He often appeared with Harry Rimmer, an itinerant evangelist and self-proclaimed scientist who wrote dozens of antievolution booklets during the decade following the Scopes trial. On at least two occasions, these two popular antievolution speakers entertained large fundamentalist audiences by debating the relative merits of the "day/age" and "gap" theories for reconciling a literal reading of the Genesis account with geological evidence of a long earth history. Neither ever wavered in his commitment to the Adam and Eve account of human creation, however.

At the same time, the Adventist science educator George McCready Price gained an increased following among fundamentalists for his creationist theory of flood geology that dispensed with any need for stretching the age of the earth beyond the under 10,000 years provided by an ultraliteral reading of the Genesis account of creation and Noah's Flood. "In the years after the Scopes trial," the historian of creationism Ronald L. Numbers noted, Price "emerged as one of the two most popular scientific authorities in fundamentalist circles, the other being Rimmer. In addition to appearing regularly in Adventist magazines, his prose frequently graced the pages of the most widely read fundamentalist periodicals." As for Rimmer, the leading conservative Christian publisher, William B. Eerdmans, reprinted his antievolution booklets in a series of books that sold over 100,000 copies during the 1940s and 1950s. Although Rimmer and Price rarely championed antievolution laws, they laid a solid foundation of antievolutionism among American fundamentalists during the post-Scopes era.[14]

Antievolutionism managed to survive and flourish even as commentators pronounced it dead and gone because its proponents focused their efforts inward, within the fundamentalist church, rather than outward, toward the general public. Beyond the church, people did not hear about Rimmer and Price during the thirties in the way they had heard about Bryan and Riley during the twenties. The leading evangelical historian, George M. Marsden, attributed this development to the Scopes trial. "It would be difficult to overestimate the impact of 'the Monkey Trial' at Dayton, Tennessee, in transforming fundamentalism," Marsden wrote. "The rural setting . . . stamped the entire movement with an indelible image. Very quickly, the conspicuous reality of the movement seemed to conform to the image thus imprinted and the strength of the movement in the centers of national life waned precipitously."[15] Fundamentalism, which began amid revivals in northern and West Coast cities, appeared increasingly associated with the rural South. The national media ceased covering its normal activities. Conservatives lost influence within mainline Protestant denominations. The string of legislative defeats for antievolution bills in northern states made further political activity outside the South seem futile. After the Scopes trial, elite American society stopped taking fundamentalists and their ideas seriously.

Indeed, *fundamentalism* became a byword in American culture as a result of the Scopes trial, and fundamentalists responded by withdrawing. They did not abandon their faith, however, but set about constructing a separate subculture with independent religious, educational, and social institutions. The historian Joel A. Carpenter traced these activities in the development of fundamentalist colleges and schools, conferences and camps, radio ministries, and missionary societies during the 1930s. The founding of Bryan College in Dayton fit the pattern perfectly. As membership in mainline Protestant associations shrank during the Great Depression, it surged ahead in most fundamentalist denominations—a phenomenon that Carpenter attributed to the role these churches played in providing "ordinary people with a compelling critique of modern society."[16]

Antievolutionism continued to feature prominently in this critique and remained a virtual tenet of Protestant fundamentalism in the United States. Rimmer, Price, and other antievolutionists spoke widely at fundamentalist churches and conferences. Their followers taught science at fundamentalist colleges and schools, which typically required all teachers and students to affirm their belief in biblical inerrancy. Bryan College twice invited Rimmer to become its president and welcomed Price to speak on campus.

Just as fundamentalists built their own religious institutions parallel to the traditional Protestant structures that shunned them, they sought to build separate institutional structures for propagating creationist scientific theories. "During the heady days of the 1920's, when their activities made frontpage headlines, creationists dreamed of converting the world; a decade later, forgotten and rejected by the establishment, they turned their energies inward and began creating an institutional base of their own," Numbers observed.[17] Price co-founded the creationist Religion and Science Association in 1935, for example, but soon left to form the stricter Deluge Geology Society. For a time, antievolutionism also found a home within the American Scientific Affiliation, a professional association of evangelical science educators created in 1941. These organizations and their journals provided an independent institutional base for creationism outside mainstream science. By the 1940s, a fundamentalist subculture had formed in the United States, with a creationist scientific establishment of its own.

Although the Scopes trial helped push fundamentalists out of mainstream American culture, they seemed almost eager to go. A separatist streak marked elements of conservative American Protestantism ever since the Pilgrims set foot on Plymouth Rock in 1620. Some distinct creationist sects, such as the Amish and Jehovah Witnesses, always isolated themselves from secular society. Others, such as the Mormons and some ultra-Orthodox Jews and Christians, tended to live in their own communities. The African-American church never had much contact with America's lily-white scientific establishment. Many strands that united under the fundamentalist banner during the early part of the century, including dispensational premillenialism and the holiness movement, had strong tendencies to renounce modern society. Their Bible told them that they were "not of this world" and that "God made foolish the wisdom of this world."[18] Bryan, Riley, and Straton prodded fundamentalists to carry their light to the world, but when the world rejected that light and martyred their champion at Dayton, the next generation of fundamentalist leaders—including John R. Rice, Carl McIntire, and Bob Jones, Sr.—called them back to separation. In the words of a popular hymn of the thirties, fundamentalists gladly sang,

> Just a few more weary days and then, I'll fly away;
> To a land where joys will never end, I'll fly away.
> . . . When I die, Hallelujah, by and by, I'll fly away."[19]

In the meantime, they felt little need to submit to the dominant culture and quietly built an ever larger and more intricate subculture of their own.

America's social elite ignored these developments for decades and institutionalized its view of the Scopes trial. Following Frederick Lewis Allen, the trial became an increasingly significant symbolic victory for liberal progress over the forces of reaction. Yet Allen dealt only with the 1920s. Political historians covering a broad sweep of modern American history faced a dilemma: Bryan stood at the center of two supposedly watershed events in American history—the populist revolt of the 1890s and the Scopes trial of the 1920s—but he had shifted sides. The same historians who deified the young Bryan of the nineties demonized the elderly Bryan of the twenties.

Richard Hofstadter, a leading American historian of the mid-twentieth century, set the tone. "Bryan decayed rapidly during his closing years. The post-war era found him identified with some of the worst tendencies in American life—prohibition, the crusade against evolution, real-estate speculation, and the Klan," Hofstadter wrote in his 1948 classic, *The American Political Tradition.* "As his political power slipped away, Bryan welcomed the opportunity to divert himself with a new crusade," he explained. "The Scopes trial, which published to the world Bryan's childish conception of religion, also reduced to the absurd his inchoate notions of democracy." In short, Hofstadter described Bryan as "a man who at sixty-five had long outlived his time." Later historians would reconstruct a more balanced picture of Bryan, showing that he never truly changed during his political career, but the Hofstadter view reigned for a generation and influenced American history textbooks even longer.[20]

The Scopes trial became a popular topic for historians during the fifties. In 1954, for example, Norman F. Furniss made it the pivotal event in his book on the fundamentalist controversy.[21] Two years later, William E. Leuchtenburg's influential book, *The Perils of Prosperity, 1914–1932*, cast antievolutionism as a peril to progress and the Scopes trial as the purgative. Ray Ginger contributed the first authoritative book-length study of the trial in 1958. Furniss and Leuchtenburg relied heavily on Allen's depiction of events at Dayton and interpretation of the outcome. For Leuchtenburg, "the campaign to preserve America as it was, to resist the forces of change, came to a head in the movement of Protestant Fundamentalism climaxed by the Scopes trial." In the end, he concluded, "The antievolutionists won the Scopes trial; yet, in a more important sense, they were defeated, overwhelmed by the tide of cosmopolitanism."[22] Ginger titled a concluding chapter, "To the Losers Belong the Spoils," and drew the lesson from Bryan's "fatal error of tactics: if a person holds irrational ideas and insists that others should accept them because of their authoritative source, he should never agree to be questioned about them."[23] In his 1955 book, *The Age of Reform: From Bryan to F.D.R.*, Hofstadter reasserted, "The pathetic postwar career of Bryan himself, once the bellwether for so many of the genuine reforms, was a perfect epitome of the collapse of rural idealism and the shabbiness of the evangelical mind."[24]

Hofstadter's collegiate American history textbook (which appeared in various editions with several co-authors beginning in 1957) presents the standard historical interpretation of the Scopes trial. In Hofstadter's work, fundamentalism appears alongside the Red Scare, the Ku Klux Klan, immigration restrictions, and Prohibition in a section on the "intolerance" that darkened the 1920s. The subsection "Fundamentalism" consists solely of a summary description of the Scopes trial. Ever since, nearly every American history survey text has lumped fundamentalism with reactionary forces during the 1920s and featured similar depictions of the Scopes trial. Many continue to perpetuate Allen's account that, as one popular textbook asserts, Scopes intentionally "lectured to his class on evolution and was arrested." Most reduce the trial to an emotional encounter between Darrow and Bryan that resulted in a decisive moral defeat for fundamentalism. Leuchtenburg's textbook called it "nineteenth-century America's last stand." Another text adopted the title "Only Yesterday" for its chapter on the twenties, concluding its account of the trial with the observation, "Darrow and company had won a signal victory by making fundamentalism henceforth the butt of ridicule." As in many of the texts, the ACLU and all of Darrow's co-counsel entirely lost their place in history.[25]

Once Riley, Straton, and other antievolution leaders associated with prosecuting the Scopes case passed from the scene, fundamentalists did little to contest the popular interpretation stamped on the trial by secular commentators and historians. Bent on separating their movement from the general culture, the next generation of fundamentalist leaders largely ignored the trial and its impact on society—a development that later, more worldly fundamentalists would come to deplore.[26] Fundamentalist students increasingly attended separate academies and colleges that, typically, did not utilize textbooks that either criticized or contradicted their faith. Most likely, only a few fundamentalists actually read what secular authors wrote about the Scopes trial, and most of them probably did not care.

Even creationist science lecturers and writers abandoned the prosecutors of John Scopes. During the late 1920s, Harry Rimmer and Arthur I. Brown defended Bryan's efforts at Dayton, but they did so less in later years.[27] The position of George McCready Price changed even more dramatically. A week before the trial, he advised Bryan to concentrate on the "utterly divisive and 'sectarian' character" of teaching evolution: "This you are capable of doing, I do not know of any one more capable." Yet Price turned against Bryan after the Commoner testified that the days of creation in Genesis represented ages of geological history. At first, Price simply commented that Bryan "really didn't know a thing about the scientific aspects of the case." By the 1940s, however, Price even surpassed secular commentators in describing the trial as a crushing defeat for fundamentalism, "which may be regarded as a turning point in the intellectual and religious history of mankind." He blamed the entire disaster on "poor Bryan, with his day-age theory of Genesis."[28] Later fundamentalist proponents of a more recent creation agreed. Price's successor at the helm of the "scientific" creationist movement, Henry M. Morris, commented, "Probably the most serious mistake made by Bryan on the stand was to insist repeatedly that he had implicit confidence in the infallibility of Scripture, but then to hedge on the geological question, relying on the day/age theory."[29] Of course, Bryan simply testified to what he and many prominent fundamentalists of his day believed. Nevertheless, late-twentieth-century fundamentalist leader Jerry Falwell maintained that Bryan "lost the respect of Fundamentalists when he subscribed to the idea of periods of time for creation rather than twenty-four hour days."[30]

During the period of fundamentalists' self-imposed isolation from the broader culture, it took threats to repeal the Tennessee antievolution statute to arouse even Bryan College stalwarts to defend the memory of Bryan's role at the trial. The first such threat came in 1935, when a 22-year-old Tennessee state representative—described in the press as a "pipe-smoking Vanderbilt law student"—offered legislation to repeal the statute. Bryan College teachers and students beseeched legislators with letters and petitions condemning the repealer. Sue Hicks, then a state representative, warned his colleagues that "repeal of the law might endanger" the college. Another lawmaker declared on the state house floor, "I believe that God looked down from high Heaven on Dayton when William Jennings Bryan was there sacrificing his blood not only in the interests of man, but in the interests of his God." A third representative maintained, "A law that was good enough for William Jennings Bryan is good enough for me." The proposal lost by a vote of 67 to 20.[31] Seventeen years later, a second effort to repeal the statute raised a similar outcry from Bryan College. Its longtime president, Judson A. Rudd, sent copies of Bryan's closing arguments to every member of the state legislature with a note stating that "the arguments advanced by

Mr. Bryan [are] as sound today as when presented twenty-five years ago."[32] Once again, the repeal effort failed.

Even though Rudd's letter defended Bryan, it suggests a further reason why midcentury fundamentalists abandoned the Commoner. "We are asking you to use your vote and influence to retain this historic and important law," Rudd wrote in this 1951 letter. "It is even more important today that we withstand the efforts of atheistic communism to deny the dignity of man and to undermine the Christian foundations of our country." To the extent that fundamentalists entered the political fray during the middle part of the century, their main concerns were with communism, which came to a peak in the early 1950s when the fundamentalist leader Carl McIntire actively supported Senator Joe McCarthy's crusade against Communist influences in America's political, education, cultural, and religious institutions.[33] From the outset, most leading fundamentalists (except Bryan) tended to lean toward the conservative end of the political spectrum, but now the movement swung hard right. Its new leaders had little inclination to defend a liberal Democratic politician such as Bryan, especially when they could blame their perceived setback at Dayton on his willingness to compromise on an ultraliteral interpretation of Genesis. Even in the early 1920s, when leading fundamentalists enlisted Bryan to aid in their fight against teaching evolution, the historian Ferenc M. Szasz observed, "it is doubtful if many of them ever voted for him. The officials of Moody Bible Institute on his death admitted that they never had." Only much later, when some evangelicals began reclaiming their heritage of social activism, did a few seek to restore Bryan's reputation.[34]

During the fifties, McCarthy-era assaults on individual liberty heightened liberal interest in fundamentalism and the Scopes trial. In particular, the sociologist of religion James Davison Hunter noted, these assaults "and the participation of conservative Protestants in them alerted the academy and the broader liberal culture to certain propensities within the conservative Protestant subculture."[35] The Scopes trial came to symbolize a moment when civil libertarians successfully stood up to majoritarian tyranny. This is apparent in Ray Ginger's 1958 book about the trial, which concludes by comparing Darrow's interrogation of Bryan with "the Senate hearings regarding Joseph R. McCarthy, where the line of questioning was weak and compromised, but the mere fact that McCarthy could be forced to answer questions at all caused millions of people to see him in a new way."[36] Similarly, Leuchtenburg's interest in the perils of prosperity during the 1920s grew out of his concern about the perils of prosperity during the 1950s—with antievolutionism standing in for anticommunism. Furthermore, Furniss began and ended his book on fundamentalism in the twenties with references to political repression of domestic dissent during the fifties.

Again, Richard Hofstadter helped set the tone. His most extensive analysis of the Scopes trial appeared in the landmark study, *Anti-Intellectualism in American Life*. "Although this book deals mainly with certain aspects of the remoter American past, it was conceived in response to the political and intellectual conditions of the 1950's," he stated at the outset. "Primarily it was McCarthyism which aroused the fear that the critical mind was at a ruinous discount in this country." Several chapters of this book discuss episodes of religious anti-intellectualism, one of which focuses on fundamentalism during the 1920s. "It was in the crusade against the teaching of evolution that the fundamentalist movement reached its climax and in the Scopes trial that it made its most determined stand," Hofstadter wrote in this chapter. Yet he described the trial as a momentous defeat for fundamentalists. "The Scopes trial, like the Army-McCarthy hearings thirty years later, brought feeling to a head and provided a dramatic purgation and resolution. After the trial was over, it was easier to see that the anti-evolution crusade was being contained," Hofstadter concluded.[37]

One significant distinction between the interpretation given the Scopes trial by historians of the 1950s and that given it by Allen and other commentators during the 1930s involves its seriousness. Both eras saw the trial as a defeat for fundamentalism, but Allen presents it primarily as a media spectacular. His account of the trial appears sandwiched between lighthearted descriptions of the mah-jongg craze and Red Grange's gridiron exploits in a chapter titled, "The Ballyhoo Years." In the shadow of McCarthyism, historians of the fifties inevitably placed it alongside the Red Scare, even though fundamentalists did not initiate or disproportionately participate in that earlier assault against alleged domestic Communists. Ballyhoo gave way to bogeymen.

Such grim fascination with the Scopes trial as a foreshadowing of McCarthyism inspired the single most influential retelling of the tale, Jerome Lawrence and Robert E. Lee's play, *Inherit the Wind*. In contrast to Allen's comic portrayal of the trial, Lawrence and Lee presented it as present-day drama. "*Inherit the Wind* does not pretend to be journalism," they wrote in their published introduction for the play, "It is not 1925. The stage directions set the time as 'Not too long ago.' It might have been yesterday. It could be tomorrow." In writing this, they did not intend to present antievolutionism as an ongoing danger—to the contrary, they perceived that threat as safely past; rather, their concern was the McCarthy-era blacklisting of writers and actors (the play opened on Broadway in 1955). "In the 1950s, Lee and his partner became very concerned with the spread of McCarthyism," a student who interviewed him reported. "Lawrence and Lee felt that McCarthyism paralleled some aspects of the Scopes trial. Lee worried, 'I was very concerned when laws were passed, when legislation limits our freedom to speak; silence is a dangerous thing.' " Tony Randall, who starred in the original Broadway cast, later wrote, "Like *The Crucible*, *Inherit the Wind* was a response to and a product of McCarthyism. In each play, the authors looked to American history for a parallel."[38]

For their model, Lawrence and Lee took Maxwell Anderson's *Winterset*, a play loosely based on the Sacco–Vanzetti case. Anderson had claimed "a poet's license to expand, develop, and interpolate, dramatize and comment," Lawrence and Lee later explained. "We asked for the same liberty . . . to allow the actuality to be the springboard for the larger drama so that the stage could thunder a meaning that wasn't pinned to a given date or a given place."[39]

The play was not history, as Lawrence and Lee stressed in their introduction. "Only a handful of phrases have been taken from the actual transcript of the famous Scopes trial. Some of the characters of the play are related to the colorful figures in that battle of giants; but they have a life and language of their own—and, therefore, names of their own." For their two starring roles, the writers chose sound-alike names: Bryan became Brady and Darrow was Drummond. The role of the Baltimore *Sun*'s H. L. Mencken was expanded to become the Baltimore *Herald*'s E. K. Hornbeck. Scopes became Cates. Tom Stewart diminished into a minor role as Tom Davenport. Malone, Hays, Neal, Rappleyea, and the ACLU disappeared from the story altogether, as did the WCFA and all the hometown prosecutors. Dayton (called Hillsboro) gained a mayor and a fire-breathing fundamentalist pastor who subjugated townspeople until Darrow came to set them free with his cool reason.

Scopes acquired a fiancée—"She is twenty-two, pretty, but not beautiful," the stage directions read, and she is the fearsome preacher's daughter. "They had to invent romance for the balcony set," Scopes later joked.[40] It may not have been accurate history, but it was brilliant theater—and it all but replaced the actual trial in the nation's memory. The play wove three fundamental changes into the story line (in addition to countless minor ones), all of which served the writers' objectives of debunking McCarthyism.

The first change involved Scopes and Dayton. Ralph Waldo Emerson once described a mob as "a society of bodies voluntarily bereaving themselves of reason." In *Inherit the Wind,* Cates becomes the innocent victim of a mob-enforced antievolution law. The stage directions begin, *"It is important to the concept of the play that the town is always visible, looming there, as much on trial as the individual defendant."* In the movie version, the town fathers haul Cates out of his classroom for teaching evolution. Limited to a few sets, the play begins with the defendant in jail explaining to his fiancée, "You know why I did it. I had the book in my hand, Hunter's *Civic Biology.* I opened it up, and read to my sophomore science class Chapter 17, Darwin's *Origin of Species.*" For innocently doing his job, Cates "is threatened with fine and imprisonment," according to the script.[41] This change provoked trial correspondent Joseph Wood Krutch. "The little town of Dayton behaved on the whole quite well," he wrote in rebuttal. "The atmosphere was so far from being sinister that it suggested a circus day." Yet, he complained, "The authors of *Inherit the Wind* made it chiefly sinister, a witchhunt of the sort we are now all too familiar with." Scopes never truly faced jail, Krutch reminded readers, and the defense actually instigated the trial. "Thus it was all in all a strange sort of witch trial," he concluded, "one in which the accused won a scholarship enabling him to attend graduate school and the only victim was the chief witness for the prosecution, poor old Bryan."[42]

Second, the writers transformed Bryan into a mindless, reactionary creature of the mob. Brady was "the biggest man in the country—next to the President, maybe," the audience heard at the outset, who "came here to find himself a stump to shout from. That's all." In the play, he assails evolution solely on narrow biblical grounds (never suggesting the broad social concerns that largely motivated Bryan) and denounces all science as "Godless," rather than the so-called false science of evolution.[43] "*Inherit the Wind* dramatically illustrates why so many Americans continue to believe in the mythical war between science and religion," Ronald Numbers later wrote. "But in doing so, it sacrifices the far more complex historical reality."[44]

On the witness stand, Brady responds even more foolishly than Bryan did at the real trial. In *Inherit the Wind,* Brady steadfastly maintains on alleged biblical authority that God created the universe in six twenty-four-hour days beginning "on the 23rd of October in the Year 4004 B.C. at—uh, at 9 A.M.!" The crowd gradually slips away from him as he babbles on, reciting the names of books in the Old Testament. "Mother. They're laughing at me, Mother!" Brady cries to his wife at the close of his testimony. "I can't stand it when they laugh at me!" At a Broadway performance of the play, the constitutional scholar Gerald Gunther became so outraged that, as he later wrote, "for the first time, I walked out of a play in disgust." He explained, "I ended up actually sympathizing with Bryan, even though I was and continue to be opposed to his ideas in the case, simply because the playwrights had drawn the character in such comic strip terms." Even though Bryan in fact opposed including a penalty provision in antievolution laws, the play ends with his character ranting against the small size of the fine imposed by the judge, then fatally collapsing in the courtroom when the now hostile crowd ignores his closing speech. *"The mighty Evolution Law explodes with a pale puff of a wet firecracker,"* the stage directions explain, just as McCarthyism itself died from ridicule.[45]

Just as Lawrence and Lee debunked Brady-Bryan in the eyes of the audience, they uplifted Drummond-Darrow. In *Inherit the Wind*, the Baltimore *Herald* engages the notorious Chicago attorney to defend Cates. Drummond makes his entrance in a *"long, ominous shadow,"* the stage directions instruct, *"hunched over, head jutting forward."* A young girl screams, "It's the Devil!" but he softens as the play proceeds. "All I want is to prevent the clock-stoppers from dumping a load of medieval nonsense in the United States Constitution," he explains at one point; "You've got to stop 'em somewhere."[46]

Drummond remains a self-proclaimed agnostic, but loses his crusading materialism. At the play's end, it is Hornbeck who delivers Darrow's famous line that Bryan "died of a busted belly" and ridicules the Commoner's fool religion. Drummond reacts with anger. "You smartaleck! You have no more right to spit on his religion than you have a right to spit on my religion! Or lack of it!" he replies. The writers have Drummond issue the liberal's McCarthy-era plea for tolerance that everyone has the "right to be wrong!" The cynical reporter then calls the defense lawyer "more religious" than Brady, and storms off the stage. Left alone in the courtroom, Drummond picks up the defendant's copy of *The Origin of Species* and the judge's Bible. After *"balancing them thoughtfully, as if his hands were scales,"* the stage directions state, the attorney *"jams them in his briefcase, side by side,"* and slowly walks off the now-empty stage.[47] "A bit of religious disinfectant is added to the agnostic legend for audiences whose evolutionary stage is not yet very high," the radical *Village Voice* sneered in its review.[48]

At the time, most published reviews of the stage and screen versions of *Inherit the Wind* criticized the writers' portrayal of the Scopes trial. "History has been not increased but almost fatally diminished," the *New Yorker* drama critic complained. "The script wildly and unjustly caricatures the fundamentalists as vicious and narrow-minded hypocrites," the *Time* magazine movie review chided, and "just as wildly and unjustly idealizes their opponents, as personified by Darrow." Reviews appearing in publications ranging from *Commonweal* and the *New York Herald Tribune* to *The New Republic* and the *Village Voice* offered similar critiques.[49]

Both the play and movie proved remarkably durable, however, despite the critics. After opening at New York's National Theater early in 1955, the stage version played for nearly three years, making it the longest-running drama then on Broadway. A touring cast took the play to major cities around the country during the late fifties. The script gained new life as a screenplay in 1960, resulting in a hit movie starring Spencer Tracy, Fredric March, and Gene Kelly. John Scopes attended its world premiere in Dayton, and thereafter promoted the movie across the country at the studio's behest. "Of course, it altered the facts of the real trial," Scopes commented, but maintained that "the film captured the emotions in the battle of words between Bryan and Darrow." Sue Hicks, the only other major participant to attend the premiere, reacted quite differently to the film. He called it "a travesty on William Jennings Bryan" and nearly purchased television time to denounce it.[50] Since its initial release, the movie has appeared continually on television and video, while the play has become a staple for community and school theatrical groups. By 1967, trial correspondent Joseph

Wood Krutch could rightly comment, "Most people who have any notions about the trial get them from the play, *Inherit the Wind*, or from the movie."[51]

All of which bothered Krutch, who had led the liberal media to Dayton. "The play was written more than a generation after the event and its atmosphere is that of the 40's and 50's, not the 20's. This makes for falsification because one of the striking facts about the whole foolish business is just that it was so characteristic of the 20's," he wrote. "That the trial could be a farce, even a farce with sinister aspects, is a tribute to the 20's when, whatever the faults and limitations of that decade, we did not play as rough as we play today." Bryan, for example, offered to pay Scopes's $100 fine; McCarthy, in contrast, destroyed careers and wrecked lives without remorse. Left unchecked, fundamentalist intolerance might have worsened but, given their natures, Bryan and other fundamentalist leaders of the twenties simply were less malign than the McCarthyites. In history classrooms, however, *Inherit the Wind* became a popular instructional tool for teaching students about the twenties. In 1994, for example, the National Center for History in Schools published instructional standards. As a means to educate high school students about changing values during the 1920s, it recommended that teachers "use selections from the Scopes trial or excerpts from *Inherit the Wind* to explain how the views of William Jennings Bryan differed from those of Clarence Darrow."[52]

As Krutch noted in 1967, "The events [at Dayton] are more a part of the folklore of liberalism than of history." The astronomer and science popularizer Carl Sagan recognized this when he observed that, even though the Scopes trial may have had little lasting impact on American culture, "the movie *Inherit the Wind* probably had a considerable national influence; it was the first time, so far as I know, that American movies made explicit the apparent contradictions and inconsistencies in the book of Genesis." Calvin College scientist Howard J. Van Till, who led the fight against antievolutionism within the evangelical church during the later part of the twentieth century, also stated that "folklore [about the Scopes trial] has had a greater impact [on American culture] than the actual historical particulars have had," but he does not so readily concede that *Inherit the Wind* monopolized that folklore. "While many members of the scientific academy might think of the Scopes trial as an episode in which Clarence Darrow artfully exposed the ignorant and narrow-minded dogmatism of North American Fundamentalism," he suggested from his experience, "many members of the conservative Christian community might think of it as an episode in which William Jennings Bryan was skillfully manipulated by a skilled but unprincipled lawyer representing an antitheistic scientific establishment."[53]

Ever since *Inherit the Wind* first appeared, conservative Christians have displayed greater interest in countering the popular impression created by it than by the trial. Creation-science leader Henry M. Morris, for example, could attribute the troubles of Bryan at Dayton to his testimony about the age of the earth but, in *Inherit the Wind*, Brady espouses a reading of Genesis every bit as literal as Morris's own. Reflecting on the problems this has caused his movement, Morris discussed a 1973 lecture tour that he gave in New Zealand. "There was a great deal of interest," he complained, "but in city after city, either during my visit or immediately afterward, the government-controlled television channels kept showing the Scopes trial motion picture, *Inherit the Wind*, over and over." Advocates of creation-science and critics of Darwinism have repeatedly attempted to explain how *Inherit the Wind* does not fairly represent their position.[54] The trial itself became, as the historian of religion Martin E. Marty later described it, "One final irrelevancy," by which he meant that it gained significance "as an event of media-mythic proportions"—that is, not for what actually occurred, but through its "acquired mythic character." For the general public since 1960, that mythic character largely came through *Inherit the Wind*.[55]

The mythic Scopes legend remained constant from *Only Yesterday* through post–World War II history textbooks to *Inherit the Wind*. The Harvard paleontologist Stephen Jay Gould summarized and criticized it as follows: "John Scopes was persecuted, Darrow rose to Scopes's defense and smite the antediluvian Bryan, and the antievolution movement then dwindled or ground to at least a temporary halt. All three parts of this story are false." Gould expressed greatest concern about the third error, which may have lulled evolutionists into a false sense of security. He noted in 1983, "sadly, any hope that the issues of the Scopes trial had been banished to the realm of nostalgic Americana have been swept aside by our current creationist resurgence."[56]

Yet the third part of this story had constituted the central lesson of the Scopes legend on which all versions concurred: The light of reason had banished religious obscurantism. In the 1930s, Frederick Lewis Allen presented the Scopes trial as a critical watershed, after which "the slow drift away from Fundamentalist certainty continued." By the fifties, antievolutionism appeared to have safely run its course. "Today the evolution controversy seems as remote as the Homeric era to intellectuals of the East," Hofstadter wrote. Lawrence and Lee left no doubts about their verdict on the Scopes trial. When the defendant asks if he won or lost, Drummond assures everyone, "You won. . . . Millions of people will say you won. They'll read in their papers tonight that you smashed a bad law. You made it a joke!" Certainly the play's actors had no doubts about this verdict. "When we did *Inherit the Wind* in 1955, the religious right was a joke, a lunatic fringe," Tony Randall later wrote. Reviewing the movie version in 1960, *The New Republic* noted, "The Monkey Trial is now a historical curiosity, and it can be made truly meaningful only by treating it as the farce that it was." While these secular interpreters of the trial contemplated the triumph of reason, however, antievolutionism continued to build within America's growing conservative Christian subculture. As Randall ruefully observes, "Sometimes we wonder if anyone ever learns anything."[57]

— CHAPTER TEN —

DISTANT ECHOES

THE SCOPES legend notwithstanding, fundamentalism had not died in Dayton—and its adherents soon reentered the political fray with many of the same concerns as their spiritual forebears of the twenties. The political landscape, however, had changed; by the late twentieth century, Americans had come to accept many of the basic notions of individual liberty championed by the ACLU during its early years. Under Chief Justice Earl Warren, the U.S. Supreme Court engrafted the ACLU view of free speech, due process, and equal protection onto the Constitution. American colleges and universities widely subscribed to the AAUP's definition of academic freedom. New Deal Congresses had enacted labor laws fully as protective of workers as those sought by Baldwin, Hays, and other ACLU founders. These legal developments made antievolution statutes seem virtually un-American by the 1960s, and led fundamentalists to seek other avenues of recourse against Darwinian teaching. Equal protection for their ideas appeared more appropriate to some fundamentalists than censoring their opponents. Furthermore, a generally acknowledged breakdown of traditional Protestant values within public education and American society left them more concerned about including creationist theories in the school curriculum than excluding evolutionary concepts from it. Their freedom and America's future demanded no less, so they thought; yet modern concepts of individual liberty made the public increasingly wary of efforts to impose religious-based rules on Americans generally. Clashes were inevitable, and recurrent.

The developing Scopes legend left antievolution statutes particularly vulnerable by the 1960s. Those laws seemed peculiar enough in the 1920s, when Bryan offered them as a means both to preserve public morality from the alleged threat of Social Darwinism and to restore neutrality on the religiously sensitive topic of human origins by teaching nothing about it. According to the Scopes legend, however, the statutes resulted from a Quixotic crusade by fundamentalists to establish their narrow religious doctrines in the classroom. Even Bryan would have regarded such an objective with concern; certainly the Warren Court that reigned in Washington would do so as well, if given the chance. The major challenge for opponents was getting the High Court to review the old statutes. Changes in American civil liberties law during the intervening forty years had all but assured their unconstitutionality.

The United States Constitution does not say much about state restrictions on individual liberty beyond the Fourteenth Amendment bar against states depriving "any person of life, liberty, or property, without due process of law." With respect to this clause, liberals of the 1920s worried mainly about conservative federal judges using it to strike down state economic regulations designed to protect workers, such as by minimum-wage and maximum-hour laws. This placed the ACLU in an awkward position when it sought to use the same clause to prevent Tennessee from imposing conditions on Scopes's employment. Taking a broad view of the matter, the liberal *New Republic* asked in 1925, "Why should the Civil Liberties Union have consented to charge the State of Tennessee with disobeying the Constitution in order legally to exonerate Mr. Scopes? They should have participated in the case, if at all, for the purpose of fastening the responsibility for vindicating Mr. Scopes, not on the Supreme Court of the United States, but on the legislature and people of Tennessee."[1] Longtime ACLU supporter Walter Lippmann took a similar position in scoring the Scopes defense. Sensitivity to this issue influenced the way in which Hays invoked the due process clause in Dayton—always stressing that it barred patently unreasonable state laws rather than those that violated any specific individual right, even freedom of speech or the establishment of religion, lest it provide authority for courts to use property rights to strike statutes.

By the 1960s, however, federal courts had long since stopped using the Fourteenth Amendment to strike down progressive state economic regulations and instead used it to void repressive state social legislation. The process began the same year as the Scopes trial, when the Supreme Court first ruled that the "liberty" protected from state infringement by the due process clause incorporated the First Amendment right of free speech. It took more than twenty years before the High Court added the establishment clause to the rights incorporated into the Fourteenth Amendment. Once it did, the Court quickly began purging well-entrenched religious practices and influences from state-supported schools. Justice Hugo Black had championed the complete incorporation of the federal Bill of Rights into the Fourteenth Amendment since his appointment to the Supreme Court during the height of New Deal disputes over the constitutionality of federal economic legislation, and later he took the lead in applying the establishment clause to public education. In 1948, Black wrote the initial decision barring religious instruction in public schools. Fourteen years later, he added the landmark opinion outlawing school prayer. In 1963, he joined in barring compulsory Bible reading from the classroom.[2] These rulings finally provided solid authority for effectively challenging antievolution statutes under the federal Constitution. The Scopes legend did the rest.

The role of science in American education also changed during this period. Cold war fears that the United States had fallen behind the Soviet Union in technology led the Congress to pass the 1958 National Defense Education Act, which pumped money into science literacy programs and encouraged the National Science Foundation to fund development of state-of-the-art science textbooks. Freed from market considerations, a team of scientists and educators working under the auspices of the Biological Sciences Curriculum Study (BSCS) produced a series of new high school biology texts that stressed evolutionary concepts. Commercial publishers rushed to keep pace. Despite scattered protests by fundamentalists, school districts throughout the country adopted the BSCS textbooks—even in the three southern states with antievolution laws.[3] No prosecutions resulted, but the new books caused some teachers to question the old laws—a few of whom took their questions to court by filing civil actions challenging the constitutionality of state laws against teaching evolution.

Two of these lawsuits played decisive roles in overturning the antievolution statutes. One began in Arkansas shortly after the Little Rock public schools adopted new textbooks in 1965. It challenged the constitutionality of that state's antievolution law, which Arkansas voters adopted by popular referendum in the wake of the Scopes trial but which local prosecutors never enforced. The state teachers' organization instituted this action, and a young biology instructor named Susan Epperson served as the nominal plaintiff. The Arkansas attorney general personally argued the state's case at trial, vainly attempting to present the statute as reasonable. He questioned the theory of human evolution by noting, among other things, that anthropologists during the preceding decade had exposed the Piltdown fossils as an elaborate hoax. Limiting the issue to Epperson's freedom to teach about various theories of origins, and cutting off specific testimony regarding any one of them, the trial judge promptly overturned the statute on federal constitutional grounds. In Tennessee a year later, Gary L. Scott threatened to challenge his state's antievolution law after losing his temporary teaching post for reportedly telling students that the Bible was "a bunch of fairy tales." His case generated headlines because it arose just as the Tennessee legislature again wrestled with repealing that law. Proponents of repeal compared Scott to Scopes as fellow victims of the statute. Indeed, the media referred to both cases as "Scopes II," and John Scopes, who recently had reemerged from obscurity after publishing his memoirs, spoke out in support of both Epperson and Scott.[4]

"I am going to review John Thomas Scopes's book," the associate editor of the *Memphis Press-Scimitar* told his boss early in 1967, "and I'd like to do it in a way that will stir up interest in getting the 42-year-old 'monkey law' repealed." The editor-in-chief agreed, and suddenly the media spotlight shifted onto the Tennessee legislature. A series of editorials and articles critical of the law ensued. Editorialists throughout the state rallied behind legislation offered by Memphis lawmakers that, as one sponsor promised, "would remove the tag 'Monkey State' from Tennessee and allow evolution to be taught."[5] The *Nashville Tennessean* tracked the bill's progress in a series of editorial cartoons showing "Col. Tennessee" in a monkey's tree, contemplating whether to climb down. The national media picked up the story, which fit neatly into the popular image of a New South trying to shed its benighted past. Fitting this image, some papers noted, the Tennessee legislature now included a few African Americans (due to federal civil rights legislation) and more urban members (as a result of reapportionment ordered by the U.S. Supreme Court). "It's been a long fight for the people of Tennessee," Scopes told reporters. "I think the people there realized that it was a bad law and would have to be repealed sooner or later. I suppose the time has come."[6]

"The upshot of the Press-Scimitar's campaign," the newspaper's owners later boasted, "was that exactly two months to the day after it started, the Tennessee legislature repealed the 'monkey law.' "[7] The lower house acted first, passing the bill by a two-to-one majority. A caged monkey bearing the sixties-era sign, "Hello Daddy-o," participated in the proceedings on the house floor. A few representatives spoke out against repeal, one of them telling his colleagues, "I've learned long ago if you try to conform to others, you will not be yourself. I care not what others say." Yet most simply wanted to free their state from any legacy of the Scopes trial. "I may be leaving," Col. Tennessee told the monkey in the next *Tennessean* cartoon. "Guess I've been a monkey long enough."[8]

The three national television networks sent camera crews to broadcast the "historic" senate vote to repeal the antievolution law. Instead, as one reporter described the scene, "The debate bogged down in public professions of faith and little discussion of the merits of the bill," ending in a sixteen-to-sixteen deadlock that temporarily preserved the status quo. "Oh, America, America," one opponent of repeal declared during the debate, "I'm a sinner and proud to testify that I believe the very word of God." A proponent countered, "I reread the Book of Genesis this morning and I do not find any conflict between Darwin's theory and the Bible." Excerpts from the debate appeared on the television news along with clips from *Inherit the Wind.* "Seems I'm still here," Col. Tennessee greeted the monkey in the next morning's paper, "Have a nut?"[9]

Opponents of the law now used Scott's threatened lawsuit, Scopes II, to pressure the senate. The national ACLU offered its assistance, as did famed defense attorney William M. Kunstler—a latter-day Clarence Darrow. Sixty Tennessee teachers and the National Science Teachers' Association joined as co-plaintiffs when Scott filed his challenge to the antievolution statute in federal court on May 15, 1967. "Nobody is asking any legislator to sacrifice any personal religious convictions by taking this law off the books," the *Tennessean* commented in its lead editorial that day. "Repeal simply means . . . that Tennessee would be saved the ordeal of another trial in which a proud state is re-

Local editorial cartoon commenting on legislative efforts to repeal the Tennessee antievolution law in 1967. (Copyrighted by the *Tennessean*, April 15, 1967. Reprinted with permission)

quired to make a monkey of itself in a court of law." Senators capitulated the next day. With network news cameras again in the chamber, they voted without debate to repeal the law. Scopes hailed the action, but one national correspondent reported "mixed feeling in Dayton about the matter." Several townspeople expressed support for the old law. "Evolution should be taught as a theory," former Scopes trial witness Harry Shelton now conceded. "Teaching it as a fact, however, is a different matter."[10]

Two weeks later, the legal issue sprang to life anew when the Arkansas Supreme Court reversed the trial judge's ruling in the Epperson case. The court did not hear oral arguments in the case or issue a formal written opinion. It simply upheld the Scopes-era law as "a valid exercise of the state's power to specify the curriculum in its public schools," and added that it "expresses no opinion on the question of whether the Act prohibits any explanation of the theory of evolution or merely prohibits teaching that theory as true."[11] The forces that had rallied around Scott now threw their support behind Epperson's further appeal. Four decades after the Scopes ruling, the ACLU finally had a decision that it could appeal to the United States Supreme Court. "The fact that the appeal now will have to be carried forward from Arkansas—rather than from Tennessee where the nonsense all started—will be readily productive of the kind of headlines that almost everybody in Arkansas seems to deplore," one Little Rock newspaper complained.[12] Before the High Court could hear the merits of the case, however, the justices had to decide to accept the appeal. Again, the Scopes legacy proved decisive.

Justice Abe Fortas took up Epperson's cause behind the scenes at the Supreme Court. After receiving the plaintiffs' petition, his young law clerk, Peter L. Zimroth, advised Fortas to "dismiss and deny" the appeal because, as the clerk wrote in a memo, "This case is simply too unreal." Zimroth explained that the statute may not bar teaching about evolution and, if it did, then prosecutors never threatened to enforce it. "Unfortunately, this case is not the proper vehicle for the Court to elevate the monkey to his proper position," he concluded. Fortas had other ideas. "Peter, maybe you're right—but I'd rather see us knock this out," he scrawled across the memo, "I'd grant or get a response." The Court went along with Fortas insofar as asking the state to respond. Arkansas's new progressive attorney general, Joseph Purcell, who had taken office since Epperson's original trial, had no special interest in the old law. He filed a perfunctory answer that did little more than assert that the statute constituted a valid exercise of state authority. "The response is as outrageous as the law which it seeks to defend," Zimroth now advised Fortas. "With you, I would like very much to strike the law down. However, I think the problems raised in my original memo are substantial. . . ." After crossing out the last phrase, he simply concluded, "I would still recommend that the court dismiss and deny." Fortas held firm, however, and the Court agreed to hear the case.[13]

The resolve of Fortas to hear the appeal probably sprang from his special interest in the Scopes case, which he experienced almost firsthand as a Tennessee public high school student during the mid-1920s. The fundamentalist–modernist controversy had swirled about him as a working-class Jewish boy growing up in the Baptist citadel of Memphis. This background certainly entered his thoughts as he considered the Epperson appeal, because his files for the case include a reply from an old friend to whom he had written about the case: "Now that the decision has been made, I should like to have a chance some day to review some of the arguments made in Breckenridge High School in 1925," the friend reminisced. "They dealt mostly with [biblical] Higher Criticism." Fortas left Tennessee for a career resembling that of Arthur Garfield Hays—including an Ivy League legal education, government service, a lucrative East Coast corporate law practice, and close ties to the ACLU in defending civil rights and liberties. Fortas dearly wanted to decide the Epperson case, and did so as one of his last majority opinions before a financial scandal forced him from the bench.[14]

Echoes of the Scopes trial resounded throughout Epperson's appeal before the Supreme Court. At the outset, Justice John M. Harlan's law clerk warned in an internal memorandum, "One objective of the Court should be to avoid a circus *à la* Scopes over this." Yet participants could hardly refrain from drawing analogies to that legendary case. The plaintiffs' principal brief to the Court closed with a dramatic reference to "the famous Scopes case" in Tennessee, and the "darkness in that jurisdiction" that followed it. The state opened its plodding written reply by appealing to the authority of the Scopes decision and closed it with extended excerpts from the Tennessee Supreme Court opinion in that case. The ACLU brief began, "The Union, having been intimately associated with *Scopes v. Tennessee* 40 years ago, when this issue first arose in the courts, looks forward to its final resolution in this case." Allusions to the Scopes case ran through the oral arguments and media coverage as well.[15]

When the justices met to discuss the case two days after oral arguments, all except Hugo Black voted to strike the law. Based on personal experience, Fortas viewed the law as an unconstitutional establishment of religion and asked the court to overturn it on that basis. According to Fortas's notes of that conference, however, most of his colleagues viewed the law as void for vagueness. "Act is too vague to stand," Chief Justice Earl Warren reportedly observed. "State has shown no need for the Act. When they prohibit teaching a doctrine, they ought to show need in terms of public order or welfare, etc." Bryan had offered such arguments long ago, as implausible as they might seem in the 1960s, but the Arkansas attorney general raised none of them. Justice William O. Douglas agreed with the chief, adding that "establishment of religion is not really in the case," presumably because all prior establishment clause rulings involved governmental actions that had the primary effect of advancing religion. Here the statute had little impact, if any. Only Harlan gave it a current effect by saying that "the law is a threat," while Black countered, "There's no case or controversy here." No one—not even Epperson's counsel under close questioning by Black during oral argument—suggested that it actually advanced religion in Arkansas. Almost alone, Fortas argued to "reverse on establishment grounds," and asked to write the Court's opinion.[16]

In the resulting opinion, Fortas set the Court's holding squarely in the context of the Scopes case, beginning and ending with references to it. He conceded that the Arkansas statute "is presently more a curiosity than a vital fact of life," yet held that it violated the establishment clause due to its original purpose. "Its antecedent, Tennessee's 'monkey law,' candidly stated its purpose," he wrote, "to make it unlawful 'to teach any theory that denies the story of Divine Creation of man as taught in the Bible, and to teach instead that man has descended from a lower order of animals.'" Never mind that this language did not appear in the Arkansas statute, he adjudged, the Tennessee law was equally on trial now. To support his analysis of the statute's historical purpose, Fortas cited the memoirs of Darrow and Scopes, a book by Richard Hofstadter, and a thirty-year-old pamphlet by the ACLU—all of which dealt with the Scopes trial rather than the Arkansas statute. Religious purpose alone became the Court's basis for striking the law.[17]

Largely as a result of the Epperson decision, having "a secular legislative purpose" became a separate test for establishment clause violations, reflecting Fortas's conviction that the clause simply must cover the Scopes situation. "In my view," the constitutional law expert Gerald Gunther later observed, "the controversy about the [Scopes] trial planted seeds of critical analysis of statutes like the Monkey Law—seeds which, decades later, bore fruit in the Supreme Court on different grounds." In a more general observation, senior legal scholar Charles Alan Wright added, "Darrow made Bryan look so foolish, as we have seen in various dramatizations of the trial, that it made the whole creationist position look foolish and made it much harder for people to insist that only creationism be taught."[18]

Justice Black could scarcely contain his frustration over the outcome of the Epperson case. In a sharply worded separate opinion, he restated his long-standing opposition to striking statutes on account of their supposed purpose. "It is simply too difficult to determine what those motives were," Black wrote. Drawing on personal experience as an Alabama politician during the antievolution crusade, the 82-year-old justice suggested an alternative purpose for the Arkansas law. Rather than favoring religious creationism, he wrote in Bryanesque fashion, "It may be instead that the people's motive was merely that it would be best to remove this controversial subject [of origins] from its schools."[19] In an apparent reply to Black, Fortas added to a later draft of his opinion the Darrowlike comment, "Arkansas' law cannot be defended as an act of religious neutrality. . . . The law's effort was confined to an attempt to blot out a particular theory because of its supposed conflict with the Biblical account, literally read."[20] Forty-three years after the Scopes trial, Black and Fortas here replayed one aspect of the debate between Bryan and Darrow—yet this time, the Darrow position clearly prevailed.

Certainly, the media played it as a long-overdue victory for Scopes. "Court Rules in a 'Scopes Case'," read the headline in one major national news magazine. *Time* led off with a reference to *Inherit the Wind*. *Life* mixed fact and fiction by reminding readers that the issue first "erupted in a glorious explosion in the tiny burg of Dayton, Tenn., where in 1925, as every student of American humor knows, Spencer

Tracy gave Fredric March the verbal thrashing of his life." A front-page article in the *New York Times* described the Epperson case as "the nation's second 'monkey trial'," but declared that it "reached a strikingly different result" from the first one.[21]

Even the seemingly decisive Epperson decision, however, failed to resolve the fundamental issues raised by the Scopes trial. This occurred in part because Fortas simplified those issues along lines suggested in the Scopes legend. In its effort to portray McCarthy-era intolerance, *Inherit the Wind* implied that antievolution laws left only biblical creationism in the classroom. Fortas carried this interpretation into the Epperson decision when he stressed that "Arkansas did not seek to excise from the curriculum of its schools and universities all discussion of the origin of man" but solely teaching about evolution. Some antievolution leaders of the 1920s might have liked to have only creationism taught, but Bryan publicly argued for the state to bar teaching evolution on the express assumption that public school teachers already could not present the biblical view.[22] By thus casting his argument as one for neutrality in education on the controversial topic of human origins, Bryan was able to gain support for antievolution laws from nonfundamentalists.

Defense counsel at Dayton did not endorse the idea of teaching both evolution and creationism in science courses. Darrow consistently debunked fundamentalist beliefs and never supported their inclusion in the curriculum. Hays and the ACLU argued for academic freedom to teach Darwinism but most likely did not consider the possibility that some teachers might want to cover creationism. Malone came the closest of anyone at Dayton to endorsing a two-view approach to teaching origins when in his great plea for tolerance he declared, "For God's sake let the children have their minds kept open—close no doors to their knowledge." Yet this came shortly after he had shouted at prosecutors, "Keep your Bible in the world of theology where it belongs and do not try to . . . put [it] into a course of science."[23] Addressing the relatively easy case of teaching only creationism as opposed to effectively ending classroom study of human origins, Fortas struck down the Arkansas antievolution law as "an attempt to blot out a particular theory from public education."[24]

Fortas clearly intended to free public schools from restrictions against teaching evolution, but his written opinion backfired when certain fundamentalists misinterpreted it as an invitation to include creationist views in public education. "In *Epperson v. Arkansas* the Supreme Court overturned a law prohibiting instruction in evolution because its primary effect was unneutral," a creationist legal strategist argued. "This unneutral primary effort [arose] . . . from an unneutral prohibition on only evolution without a similar proscription on *Genesis*."[25] Following such reasoning, some fundamentalists called for balancing instruction in evolution with creationist teaching as a supposedly constitutional alternative to excluding any one theory. Fortas may have thought that the earlier Supreme Court ruling barring religious instruction in public schools adequately covered this situation, but he did not anticipate the tenacity of fundamentalists who believed that scientific support existed for their creationist beliefs. Bills and resolutions mandating equal time or balanced treatment for creationism soon began appearing before state legislatures and local school boards throughout the nation. Proponents turned the Scopes legend to their benefit by widely quoting a fictitious statement attributed to Darrow at Dayton, "It is 'bigotry for public schools to teach only one theory of origins.' "[26]

Of course, the force of this movement sprang from the vast number of Americans who hold creationist views, and not from any encouragement given it by either the Epperson opinion or the Scopes legend. "Debate over the origin of man is as alive today as it was at the time of the famous Scopes trial in 1925," pollster George Gallup reported, on the basis of a 1982 public opinion survey, "with the public now about evenly divided between those who believe in the biblical account of creation and those who believe in either a strict interpretation of evolution or an evolutionary process involving God." This and other polls consistently found over 80 percent support for including creationist theories in the curriculum.[27]

On the strength of such sentiments, three states adopted laws mandating creationist instruction in public schools before the Supreme Court stepped in to stem the tide. In 1974, Tennessee mandated "an equal amount of emphasis" in biology textbooks for alternative theories of origins, expressly including the Genesis account. Seven years later, Arkansas and Louisiana enacted laws requiring "balanced treatment" in biology instruction for "creation-science": the Arkansas act linked this so-called science to the study of a biblically inspired list of creation events, such as a worldwide flood, while the Louisiana statute defined it as "scientific evidence for creation and inferences from those scientific evidences."[28] These three laws fell in separate lawsuits, and the media compared each of them to the Scopes case.

The Scopes legacy did more than merely influence media coverage of these cases; it shaped their very tone and timber. Drawn by the Scopes connection, the ACLU led the fight against all three statutes, with prominent New York counsel serving as their agents in the latter two cases. "It is a strange feeling," the ACLU's 97-year-old founding director Roger Baldwin commented upon passage of the Louisiana statute, "here's where I came in [with Scopes], and here's where the ACLU goes out to another battle to defend the same principles of freedom."[29] Challengers stressed the Scopes connection in all three lawsuits because it highlighted the religious purposes underlying the statutes and thereby provided a ready basis for striking them down. The first two statutes obviously violated establishment clause principles by expressly mandating public school instruction in biblical doctrines, and federal courts quickly disposed of them. The Louisiana statute simply called for teaching about scientific evidence for creation, however, and its defenders maintained that such teaching would not constitute religious instruction. Here, the Scopes legacy helped the challengers to prevail.

"The case comes to us against a historical background that cannot be denied or ignored," a federal appeals-court panel noted in its analysis of the Louisiana statute. "The Act continues the battle William Jennings Bryan carried to his grave. The Act's intended effect is to discredit evolution by counterbalancing its teaching at every turn with the teaching of creationism, a religious belief. The statute therefore is a law respecting a particular religious belief . . . and thus is unconstitutional." A bare majority of the circuit's fifteen judges affirmed this ruling on review, but seven dissented—and tried to turn the Scopes legacy inside out. "The *Scopes* court upheld William Jennings Bryan's view that states could constitutionally forbid teaching the scientific evidence for the theory of evolution," Judge Thomas Gibb Gee wrote for the dissenters. "By requiring that the whole truth be taught, Louisiana aligned itself with Darrow; striking down this requirement, the panel holding aligns us with Bryan." Both sides thus claimed the moral high ground that was by then almost universally associated with the Scopes defense.[30]

The battle over the Scopes legacy continued when the Supreme Court agreed to review the Louisiana statute. "We need not be blinded in this case to the legislature's preeminent religious purpose in enacting this statute," Justice William J. Brennan, Jr., wrote for the majority. He then referred "to the Tennessee statute that was the focus of the celebrated *Scopes* trial in 1925" as an antecedent for the Louisiana law. Writing for the dissent, however, Justice Antonin Scalia offered quite a different view of the Scopes precedent. "The people of Louisiana," he contended, "including those who are Christian fundamentalists, are quite entitled, as a secular matter, to have whatever scientific evidence there may be against evolution presented in their schools, just as Mr. Scopes was entitled to present whatever scientific evidence there was for it."[31] These clashing applications of the Scopes legend illustrate its broad appeal as folklore. Brennan could just as easily invoke it to support freedom from religious establishment as Scalia could use it to support academic freedom to teach alternative theories.

Some fundamentalists already have adopted the latter approach. When state or local education officials seek to follow the Supreme Court decisions on religious instruction in public schools by stifling conservative Christian teachers from presenting evidence for creationism in science classrooms (as happens with increasing frequency), antievolutionists often liken it to the alleged persecution of John Scopes. Courts readily dismiss the analogy by reasoning that Scopes wanted to teach a scientific theory while the others wanted to present their religious beliefs. This does not satisfy fundamentalists, however, who view their beliefs as truer than any scientific theory, because for them religion (and not science) is founded on personal experiences and relationships.

In a thoughtful discussion about such a case that arose in California during the early 1990s, the Yale law professor Stephen L. Carter concluded that the issue ultimately involves questions of epistemology. Who does have "the right," he asked, to decide what gets taught as science in the public schools? Creationist parents and teachers, based on their relatively subjective religious beliefs, or professional scientists and educators, based on their relatively objective scientific theories? "The rhetorical case against the creationist parents rests not merely or mostly on arcane questions of constitutional interpretation," Carter observes, "the case rests on the sense that they themselves are wrong to rely on their sacred texts to discover truths about the world."[32] Darrow fully realized this at Dayton, and used his defense of Scopes to challenge fundamentalist beliefs. To the extent that lawyers defending the evolutionist position in later lawsuits appeal narrowly to constitutional interpretation, fundamentalist beliefs remain unchallenged.

Certainly the court decisions since the Scopes case have not slowed the spread of creationism. Instead, they have encouraged fundamentalists to abandon evolution-teaching public education for creation-affirming church or home schooling. This relatively new development built on the earlier movement for separate fundamentalist colleges that went at least as far back as the fundamentalist–modernist controversy and gained momentum after the Scopes trial. Concern over teaching evolution contributed to both developments. In his foreword to a 1974 biology textbook written for fundamentalist high schools, for example, the creationist leader Henry M. Morris attributes "the widespread movement in recent years toward the establishment of new private Christian schools" to the perception among fundamentalist pastors and parents that "a nontheistic religion of secular evolutionary humanism has become, for all practical purposes, the official state religion promoted in the public schools."[33] His text offers a markedly different theology for the science classroom.

Not all conservative Christians reacted to the Scopes legacy with such defiance, however, especially after a self-proclaimed "new evangelical" strain of American Protestantism emerged following the Second World War under the inspiration of William Bell Riley's hand-picked successor, the evangelist Billy Graham. In his public ministry, Graham ignored the Scopes trial and antievolutionism. In 1954, he endorsed *The Christian View of Science and Scripture*, a new book by the Baptist theologian Bernard Ramm that sought to reconcile conservative Christians to modern science by interpreting the Genesis account as a pictorial depiction of progressive creationism spanning eons. Ramm's influential book, which cleared a path to the serious study of science for a generation of evangelical college students, dismissed "Bryan's miseries at the Scopes trial," as Ramm called them, as part of a "sordid history" that "we will not trace."[34] This approach fit Graham's objective of resurrecting a biblically orthodox creed free from the cultural baggage that made fundamentalism unacceptable to most educated Americans. Mindful of the ridicule heaped on Bryan for his testimony at Dayton, scholars within the new evangelical movement typically view militant antievolutionism as deadweight to be cast off.

Many other American Christians feel even less direct impact from the Scopes legacy than evangelicals. Modernists and mainline Protestants typically share the common culture's reaction to the trial and legend. Despite their traditionalism, American Catholics did not join Bryan's antievolution crusade, in part because they already had their own parochial schools and colleges, which left them in the position of spectators to the Dayton trial and its aftermath. Rooted in a historic faith adaptable enough to accept theistic evolution, Roman Catholics sat out this culture clash. Yet the issue will never wholly disappear so long as fundamentalists continue to object to teaching evolution, which they persist in seeing as damnable indoctrination in a naturalistic worldview that undermines belief in God.

Certainly the Scopes legacy clings fast to Tennessee, where most people still profess the Christian faith and most Christians lean toward fundamentalism. Republicans targeted that traditionally Democratic state during the 1994 elections, with strong support from conservative Christian political forces. In an attempt to survive the onslaught, the state's senior Democratic U.S. senator went so far as to prepare a television commercial touting his support for school prayer, but to no avail. Republicans swept into power throughout Tennessee, and new legislation to restrict teaching evolution in public schools soon appeared in the state senate with the support of fundamentalist groups and individuals. About the same time, the Alabama board of education ordered that new biology textbooks carry a disclaimer identifying evolution as "a controversial theory . . . , not fact," and the Georgia house of representatives passed a measure facilitating instruction in creationism. "Yet it's the Tennessee debate that has helped put the issue on the national stage," *USA Today* reported. "It was in Tennessee in 1925 that the two sides squared off in Scopes' epic trial." The feature article discussed the 70-year-old trial at length, and included pictures of Darrow, Bryan, and Scopes.[35]

Largely due to the Scopes connection, the new legislation drew international attention. "Seventy years after John Scopes was convicted of teaching evolution in Dayton, Tenn., the State Legislature here is considering permitting school boards to dismiss teachers who present evolution as fact rather than a theory of human origin," began a front-page article from Nashville in the Sunday *New York Times*.[36] The British Broadcasting Corporation sent a camera crew to cover the story, complete with interviews in Dayton. Some American network news accounts featured clips from *Inherit the Wind.* Newspaper articles inevitably dwelt on the Scopes trial. Amid a flurry of hostile media coverage, the senate education committee approved the proposal by an eight-to-one vote, and sent it on to the full senate, which debated the two-sentence bill for three days. "Coming more than 70 years after Tennessee's 1925 anti-evolution law was held up to international ridicule during the Scopes Monkey Trial, the bill again has brought national attention to Tennessee's ongoing debate of how to teach the origins of life on Earth. Cameras and reporters jammed into the Senate for the debate," the Memphis *Commercial Appeal* reported.[37]

Opponents dubbed the bill "Scopes II" and "Son of Scopes." They devoted more effort to warning of its public-relations impact than to defending the theory of evolution. "This echo of the 1925 law that led to the Scopes monkey trial," the *Nashville Banner* commented, "can't help but make the state look bad." The ACLU vowed to challenge the law in court, with its Nashville director warning, "I have already had several calls from teachers who are willing and interested in being plaintiffs, people who are interested in being the next John Scopes." Finally, the senate's presiding officer and senior member declared, "I can't vote for this bill, but I don't want anybody to think I don't know God," and the bill failed by a vote of twenty to thirteen. Observers credited the Scopes legacy for the defeat.[38]

The legislation evoked mixed reactions in Dayton. "I believe if they had the trial again today it would turn out about the same way," Harry Shelton had commented a few years earlier, although he grudgingly conceded, "Now they permit the teaching of evolution in most schools—as long as you teach it as a theory and not as a fact." Another former student called the new legislation "Silly, silly," and Fred Robinson's now elderly daughter added, "It's a lot of hooey." Teachers at the new regional high school keep quiet about the proposal at the request of their principal. The town's population has tripled since 1925, spurred by a new furniture factory and better roads to Chattanooga. Memories of the trial draw tourists, too, with a Scopes Trial Museum in the old courthouse and an annual Scopes Festival featuring dramatic reenactments in the courtroom. The local newspaper editor likes the proposed new statutory limits on the teaching of evolution. "To my knowledge, it's never been proven, even when we put on the trial here," he noted. From the hill above town, Bryan College's creationist biology professor agreed, adding that the bill "strikes a very profound chord in an awful lot of people." In addition, these people—fundamentalists mostly—continue to read and hear arguments (much like those once made by Bryan) that challenge the scientific authority of Darwinism. With Bryan College faculty overseeing the town's portrayal of the Scopes trial, the Commoner and his ideas still get a fair hearing in Dayton.[39]

The deeply entrenched Scopes legend continues to dominate impressions of the trial elsewhere. Even in Nashville, the morning newspaper dubbed debate on the 1996 legislation as "Inherit the Wind: The State Sequel."[40] One week after the bill's defeat, Tony Randall's production company revived Lawrence and Lee's play on Broadway, with the character representing Bryan appearing fatter and more disreputable than before. Theater critics hailed the play as pertinent and timely. "We still have the creationists versus the evolutionists," a reviewer on public television commented, and pointed to the new anti-evolution bill "in, yes, the state of Tennessee." Whereas its review of the original Broadway production criticized the script's "overall lack of tension" and "clinical quality," the *New York Times* now praised the text's "dramatic life." The critic explained, "Here was a headline-making heavyweight bout between the rational thought of a newly rational age and old-fashioned Christian fundamentalism, which was deemed to be on its last legs, though today it's alive and well and called Creationism." *The New Yorker*, which originally scorned the play as "a much too elementary study in black and white," now lauded it as "a thoughtful, powerful explication of religious and political issues that we still haven't figured out." A sign in the theater lobby quoted 1996 presidential candidate Patrick Buchanan's comments in support of the Tennessee bill.[41]

These changing responses help account for the enduring public interest in both the play and the trial. To "intellectuals" of the 1950s, as Hofstadter noted, the Scopes trial seemed "as remote as the Homeric era," and some of them criticized the play's simplistic presentation of America's debate over science and religion. Such critics typically accepted a scientific explanation for human origins and assumed that virtually all thinking Americans did so too, even those who believed in

God. Certainly *Inherit the Wind* grossly simplified the trial, yet regardless of their position on the issue, many Americans perceive the relationship between science and religion in just such simple terms: either Darwin or the Bible was true. Hofstadter recognized this. "The play seemed on Broadway more like a quaint period piece than a stirring call for freedom of thought," he observed. "But when the road company took the play to a small town in Montana, a member of the audience rose and shouted 'Amen!' at one of the speeches of the character representing Bryan."[42]

As the amens for creationism have increased in both number and volume over the years since 1955, secular critics have tended to revise their views of the play and the trial. Even aloof intellectuals have come to realize that a vast number of Americans still believe in the Bible and accept it as authoritative on matters of science. Moreover, if people accept the biblical account of special creation over the scientific theory of organic evolution, which is, after all, one of the core theories of modern biology, then they most likely defer to biblical authority on other matters of public and private concern. For Americans who do not share this religious viewpoint and who fear that fundamentalists constitute the majority in some places, concerns about the defense of individual liberty under a government by the people seem all too familiar. The character representing Darrow in *Inherit the Wind* might just as well be standing in the doorway of their bedrooms as that of a small town's schoolhouse—blocking the entrance of frenzied townspeople, and turning them aside by debunking their overzealous leader. The original Broadway cast did not take fundamentalist politicians seriously, Tony Randall observed shortly after the play's revival in 1996, "but America has moved so far to the right, that they are now close to the center."[43]

Although Lawrence and Lee's dramatic plea for tolerance originally may have been targeted against the McCarthyites, with fundamentalists standing in as straw men, the straw men have proven to be more durable than the intended targets—and the threat to individual liberty that they symbolize has become increasingly ominous for some Americans as the power of government has grown over the ensuing years. Indeed, the issues raised by the Scopes trial and legend endure precisely because they embody the characteristically American struggle between individual liberty and majoritarian democracy, and cast it in the timeless debate over science and religion. For twentieth-century Americans, the Scopes trial has become both the yardstick by which the former battle is measured and the glass through which the latter debate is seen. In its 1996 review of *Inherit the Wind*, the *New York Times* described the original courtroom confrontation as "one of the most colorful and briefly riveting of the *trials of the century* that seemed to be especially abundant in the sensation-loving 1920's."[44] Dozens of prosecutions have received such a designation over the years, but only the Scopes trial fully lives up to its billing by continuing to echo through the century.

NOTES

Throughout the notes, page numbers of newspaper articles refer to the first page of the article, and the following sources are identified in short form:

ACLU Archives: American Civil Liberties Union Archives, Princeton University Libraries, Princeton, N.J.
Bryan Papers: William Jennings Bryan Papers, Library of Congress, Washington, D.C.
Darrow Papers: Clarence Darrow Papers, Library of Congress, Washington, D.C.
Fortas Papers: Abe Fortas Papers, Princeton University Libraries, Princeton, N.J.
Hicks Papers: Judge Sue K. Hicks Papers, University of Tennessee Libraries, Knoxville, Tenn.
Mims Papers: Edwin Mims Papers, Vanderbilt University Libraries, Nashville, Tenn.
Peay Papers: Official Papers of Governor Austin Peay, Tennessee State Archives, Nashville, Tenn.
Transcript: *The World's Most Famous Court Case: Tennessee Evolution Case.* Dayton: Bryan College, 1990.

Introduction

1. All quotations from the trial are taken from Transcript, 284–304.
2. William Jennings Bryan, *In His Image* (New York: Revell, 1922), 13.
3. Transcript, 323.
4. William Jennings Bryan, *Is the Bible True?* (Nashville: private printing, 1923), 10.

5. Henry Fairfield Osborn, "Evolution and Religion," *New York Times*, 5 March 1922, sec. 7, p. 2.

6. Transcript, 236–38, 244–45, 277–78.

7. "The Scopes Trial," *Chicago Tribune*, 17 July 1925, p. 8.

Chapter One. Digging Up Controversy

1. Charles Dawson and Arthur Smith Woodward, "On the Discovery of Palaeolithic Human Skull and Mandible," *Quarterly Journal of the Geological Society of London* 69 (1913), 117.

2. For a scientific description of these fossil remains from the time, see Arthur Keith, *The Antiquity of Man* (London: Williams & Norgate, 1915), 497–511; and for a somewhat later description by an expert witness for the defense in the Scopes trial, see Kirtley F. Mather, *Old Mother Earth* (Cambridge: Harvard University Press, 1929), 52–55.

3. Dawson and Woodward, "Discovery," 133–35, 139.

4. Boyd Dawkins, in discussion following ibid., 148–49.

5. "Paleolithic Skull Is a Missing Link; Bones Probably Those of a Direct Ancestor of Modern Man," *New York Times*, 19 December 1912, p. 6.

6. "Man Had Reason Before He Spoke," *New York Times*, 20 December 1912, p. 6.

7. For example, "Exhibit Skull Believed Oldest Ever Discovered," *Chicago Tribune*, 20 December 1912, p. 9.

8. "Darwin Theory Proved True; English Scientists Say the Skull Found in Sussex Establishes Human Descent from Apes," *New York Times*, 22 December 1912, p. C1.

9. "Simian Man," *New York Times*, 22 December 1912, p. 12.

10. See, e.g., Edward Hitchcock and Charles H. Hitchcock, *Elementary Geology* (New York: Ivison, 1860), 377–93; James D. Dana, *Manual of Geology*, 2d ed. (New York: Ivison, 1895), 767–70.

11. C. I. Scofield, ed., *Scofield Reference Bible* (New York: Oxford University Press, 1909), 3*n*2, 4*nn*1,2.

12. Ronald L. Numbers, *The Creationists: The Evolution of Scientific Creationism* (New York: Knopf, 1992), 7.

13. George William Hunter, *A Civic Biology* (New York: American, 1914), 253.

14. Charles Darwin, *The Origin of Species by Charles Darwin: A Variorum Text*, ed. Morse Pechham (Philadelphia: University of Pennsylvania Press, 1959), 747. Darwin goes on to add that Lamarckian-type factors might also cause variation.

15. Charles Darwin to Asa Gray, 22 May 1860, in Francis Darwin, ed., *Life and Letters of Charles Darwin*, vol. 2 (New York: Appleton, 1896), 105.

16. T. H. Huxley to Bishop of Ripon, 19 June 1887, in Leonard Huxley, *Life and Letters of Thomas Henry Huxley*, vol. 2 (New York: Appleton, 1901), 173.

17. T. H. Huxley to Charles Darwin, 23 November 1859, in ibid., vol. 1, 189.

18. T. H. Huxley to Charles Kingsley, 30 April 1863, in ibid., 52.

19. Charles Hodge, *What is Darwinism?* (New York: Scribner's, 1874), 11, 173.

20. Asa Gray, *Natural Selection and Religion: Two Lectures Delivered to the Theological School of Yale College* (New York: Scribner's, 1880), 68–69.

21. "Introduction," *American Naturalist* 1 (1867), 2.

22. Joseph LeConte, *Evolution and Its Relation to Religious Thought* (New York: Appleton, 1891), 258, 301.

23. Clarence King, "Catastrophism and Evolution," *American Naturalist* 2 (1877), 470.

24. E. D. Cope, *Theology of Evolution: A Lecture* (Philadelphia: Arnold, 1887), 31.

25. For example, at the time of the Scopes trial, the antievolution leader William Bell Riley lauded LeConte as an example of a scientist who believed that "there must be an infinite Creator back of nature." W. B. Riley, "Should Evolution Be Taught in Tax Supported Schools?" (1928), in Ronald L. Numbers, ed., *Creation–Evolution Debates* (New York: Garland, 1995), 371.

26. Peter J. Bowler, *Evolution: The History of an Idea* (Berkeley: University of California Press, 1984), 233.

27. Vernon L. Kellogg, *Darwinism To-Day* (New York: Holt, 1907), 5.

28. Julian Huxley, *Evolution: The Modern Synthesis* (London: Chatto and Windus, 1968); William Jennings Bryan, "The Prince of Peace," in William Jennings Bryan, ed., *Speeches of William Jennings Bryan*, vol. 2 (New York: Funk & Wagnalls, 1909), 266–67.

29. A. H. Strong, *Systematic Theology*, vol. 2 (Westwood: Revell, 1907), 473.

30. B. B. Warfield, *Biblical and Theological Studies* (New York: Scribner's, 1911), 238.

31. James Orr, *God's Image of Man* (London: Hodder & Stoughton, 1904), 96.

32. James Orr, "Science and the Christian Faith," in *The Fundamentals: A Testimony to the Truth* 7 (Chicago: Testimony, [1905–15]), 102–3 (emphasis in original).

33. John William Draper, *History of the Conflict Between Religion and Science* (New York: Appleton, 1874), vi.

34. Andrew Dickson White, *The Warfare of Science* (London: King, 1876), 7.

35. Orr, "Science and Faith," 89. The historian George M. Marsden wrote about Draper and White, "Though dubious reconstructions of the evidence (usually ignoring, for instance, that most of the debate about science had been debates among Christians) they suggested that the intellectual life of the past several centuries had been dominated by the conflict between advocates of religious based obscurantism and enlightened champions of value-free scientific truth." George M. Marsden, *Understanding Fundamentalism and Evangelicalism* (Grand Rapids, Mich.: Eerdmans, 1991), 139–40.

36. Edwin Mims, "Modern Education and Religion," manuscript of address to the Association of American Colleges, in Mims Papers.

37. A. W. Benn and F. R. Tennant, quoted in James R. Moore, *The Post-Darwinian Controversies* (Cambridge: Cambridge University Press, 1979), 41, 47.

38. Arthur Keith, *Concerning Man's Origin* (London: Watts, 1927), 41 (reprint of essay first published in the *Rationalist Press Association Annual* for 1922).

39. Clarence Darrow, *The Story of My Life* (New York: Grosset, 1932), 250.

40. Clarence Darrow, quoted in Kevin Tierney, *Darrow: A Biography* (New York: Croswell, 1979), 85; Arthur Weinberg and Lila Weinberg, *Clarence Darrow: A Sentimental Rebel* (New York: Putnam's, 1980), 42.

41. "Malone Glad Trial Starts on Friday," *Chattanooga Times*, 19 July 1925, p. 2; Arthur Garfield Hays, "The Strategy of the Scopes Defense," *Nation*, 5 August 1925, p. 158.

42. W. C. Curtis, "The Evolution Controversy," in Jerry R. Tompkins, ed., *D-Days at Dayton: Reflections on the Scopes Trial* (Baton Rouge: Louisiana State University Press, 1965), 75.

43. Moore, *Post-Darwinian Controversies*, 73.

44. Asa Gray, *The Elements of Botany for Beginners and Schools* (New York: Ivison, 1887), 177.

45. Joseph LeConte, *A Compend of Geology* (New York: Appleton, 1884), 242–82, 313–90.

46. James Edward Peabody and Arthur Ellsworth Hunt, *Elementary Biology: Plants* (New York: Macmillan, 1912), 118.

47. Clifton F. Hodges and Jean Dawson, *Civic Biology* (Boston: Ginn, 1918), 331–35.

48. George William Hunter, *A Civic Biology: Presented in Problems* (New York: American, 1914), 194–96, 405.

49. Statistics from U.S. Department of Commerce, Bureau of Census, *Historical Statistics of the United States*, vol. 1 (Washington, D.C.: Government Printing Office, 1975), 16, 368–69; Tennessee Department of Education, *Annual Report for 1925* (Nashville: Ambrose, 1925), 165.

50. Austin Peay, "The Second Inaugural—1925," in *Austin Peay, Governor of Tennessee, 1923–29: A Collection of State Papers and Public Addresses* (Kingsport, Tenn.: Southern, 1929), 211.

51. Bettye J. Broyles, *Churches and Schools in Rhea County, Tennessee* (Dayton: Rhea County Historical and Genealogical Society, 1992), 258.

52. Thomas Hunt Morgan, *A Critique of the Theory of Evolution* (Princeton, N.J.: Princeton University Press, 1916), 194.

53. Thomas Hunt Morgan, *The Scientific Basis of Evolution* (New York: Norton, 1932), 109–10.

54. For example, Curtis asserted in his affidavit as expert witness for the defense at the Scopes trial, "The modern science of genetics is beginning to solve the problem of how evolution takes place, although this question is one of ex-

treme difficulty." The antievolutionist Harold W. Clark, who taught science at a small church college, sought to refute Curtis on this point but in doing so acknowledged that modern discoveries in genetics were reviving the idea that slight, random variations could account for evolution. Harold W. Clark, "Back to Creationism," in Ronald L. Numbers, ed., *The Early Writings of Harold W. Clark and Frank Lewis Marsh* (New York: Garland, 1995), 100.

55. Albert Edward Wiggam, *The Next Age of Man* (Indianapolis: Bobs-Merrill, 1927), 43.

56. William Jennings Bryan, "God and Evolution," *New York Times*, 26 February 1922, sec. 7, p. 1 and sec. 7, p. 11.

57. Henry Fairfield Osborn, "Evolution and Religion," *New York Times*, 5 March 1922, sec. 7, p. 2. In the same year, Princeton University naturalist Edwin Conklin issued a similar public attack on antievolutionism in which he charged, "Uncertainty among scientists as to cause of evolution has been interpreted by many non-scientific persons as throwing doubt upon its truth." Edwin G. Conklin, *Evolution and the Bible* (Chicago: American Institute of Sacred Literature, 1922), 3. Both Osborn and Conklin were liberal Christians, and Conklin's defense of teaching evolution appeared in a series of modernist religious tracts.

58. Thomas Hunt Morgan, *What Is Darwinism?* (New York: Norton, 1927), viii–ix (reprint of earlier popular article) (emphasis in original).

59. Bryan, "Prince of Peace," 269.

60. George W. Hunter and Walter G. Whitaman, *Science in Our World of Progress* (New York: American, 1935), 486.

61. Hunter, *Civic Biology*, 263.

62. William Jennings Bryan, *In His Image* (New York: Revell, 1922), 108; Transcript, 333–36.

63. Billy Sunday, "Historical Fabric of Christ's Life Nothing Without Miracles," *Commercial Appeal* (Memphis), 7 February 1925, p. 13.

64. Albert Edward Wiggam, *The New Decalogue of Science* (Indianapolis: Bobbs-Merrill, 1922), 105. In his closing argument for the Scopes trial, as part of his attack on evolutionary theory, Bryan expressly denounced this book and the eugenic ideas that it promoted.

65. Wiggam, *Next Age of Man*, 45 (emphasis in original).

66. Raymond A. Dart, *Adventures with the Missing Link* (New York: Harper, 1959), 5.

67. Raymond A. Dart, "Australopithecus africanus: The Man-Ape of South Africa," *Nature* 115 (1925), 198.

68. Robert Broon, "Some Notes on the Taungs Skull," *Nature* 115 (1925), 571.

69. Raymond A. Dart to Arthur Keith, 26 February 1925, in Frank Spencer, ed., *The Piltdown Papers, 1908–1955* (London: Oxford University Press, 1990), 160 (emphasis in original).

70. Dart, *Adventures with the Missing Link*, 7.

71. Dart, "Australopithecus," 198–99.

72. Bryan, "Prince of Peace," 269.

73. Dart, *Adventures with the Missing Link*, 38–40 (includes quotations from newspapers and magazines); William Jennings Bryan, "Mr. Bryan Speaks to Darwin," *Forum* 76 (1925), 102–3. At about the same time, antievolution science lecturer Harry Rimmer asserted that the Piltdown hominid "is made up of plaster of Paris and imagination," while William Bell Riley referred to it as "imaginatively created." Harry Rimmer, "Monkeyshines: Fakes, Fables, Facts Concerning Evolution," in Edward B. Davis, ed., *The Antievolution Pamphlets of Harry Rimmer* (New York: Garland, 1995), 427; W. B. Riley, "Evolution—A False Philosophy," in William Vance Trollinger, Jr., ed., *The Antievolution Pamphlets of William Bell Riley* (New York: Garland, 1995), 101.

Chapter Two. Government by the People

1. William Jennings Bryan, "God and Evolution," *New York Times*, 26 February 1922, sec. 7, p. 1.

2. Henry Fairfield Osborn, "Evolution and Religion," *New York Times*, 5 March 1922, sec. 7, p. 14.

3. John Roach Straton, "In the Negative," in John Roach Straton and Charles Francis Potter, *Evolution Versus Creation* (1924), reprinted in Ronald L. Numbers, ed., *Creation–Evolution Debates* (New York: Garland, 1995), 88–89. The fundamentalist leader William Bell Riley later took a similar position; see W. B. Riley, *Evolution—A False Philosophy*, reprinted in William Vance Trollinger, Jr., ed., *The Antievolution Pamphlets of William Bell Riley* (New York: Garland, 1995), 111–12.

4. George McCready Price, *The Phantom of Organic Evolution* (New York: Revell, 1924), 110–11. For a representative example of Osborn's dating of these fossils, see Henry Fairfield Osborn, *Evolution in Religion and Education* (New York: Scribner's, 1926), 146.

5. William Jennings Bryan, "Speech to the West Virginia State Legislature," in William Jennings Bryan, *Orthodox Christianity Versus Modernism* (New York: Revell, 1923), 37.

6. William Jennings Bryan, "The Prince of Peace," in William Jennings Bryan, ed., *Speeches of William Jennings Bryan* (New York: Funk & Wagnalls, 1909), 267.

7. A. C. Dixon and R. A. Torrey, quoted in Ronald L. Numbers, *The Creationists: The Evolution of Scientific Creationism* (New York: Knopf, 1992), 39.

8. Shailer Mathews, "Modernism as Evangelical Christianity," in Mark A. Noll et al., eds., *Eerdmans' Handbook to Christianity in America* (Grand Rapids, Mich.: Eerdmans', 1983), 379.

9. "Editorial," *Our Hope* 25 (July 1918), 49.

10. George M. Marsden, *Fundamentalism and American Culture: The Shaping of Twentieth-Century Evangelicalism, 1870–1925* (New York: Oxford University Press, 1980), 149.

11. Ibid., 157–58.

12. William Bell Riley, *Message to the Metropolis* (Chicago: Winona, 1906), 24–48, 165–95, 224–27 (quote on 48).

13. Transcribed proceedings of the WCFA conference were published as *God Hath Spoken* (Philadelphia: Bible Conference Committee, 1919), 27, 221, 441.

14. [Curtis Lee Laws], "Convention Side Lights," *Watchman-Examiner* 8 (1920), 834.

15. William Bell Riley, quoted in Ferenc Morton Szasz, *The Divided Mind of Protestant America, 1880–1930* (Tuscaloosa: University of Alabama Press, 1982), 107.

16. William Jennings Bryan, "Applied Christianity," *The Commoner,* May 1919, p. 12.

17. William Jennings Bryan, *America and the European War* (New York: Emergency Peace Federation, 1917), 14; William Jennings Bryan, quoted in Jonathan Daniels, *The Wilson Era: Years of Peace, 1910–17* (Chapel Hill: University of North Carolina Press, 1944), 428.

18. William Jennings Bryan, quoted in Lawrence W. Levine, *Defender of the Faith: William Jennings Bryan, The Last Decade, 1915–1925* (New York: Oxford University Press, 1965), 274.

19. Levine, *Defender of the Faith*, vii.

20. Bryan, "Prince of Peace," 266–68.

21. Ibid., 268–69.

22. Vernon Kellogg, *Headquarters Nights* (Boston: Atlantic, 1917), 22, 28.

23. William Jennings Bryan, *Shall Christianity Remain Christian? Seven Questions in Dispute* (New York: Revell, 1924), 146.

24. James H. Leuba, *The Belief in God and Immortality* (Boston: Sherman, French, 1916), 203, 213, 254.

25. William Jennings Bryan, *In His Image* (New York: Revell, 1922), 118.

26. Ibid., 120.

27. William Jennings Bryan and Mary Baird Bryan, *The Memoirs of William Jennings Bryan* (Philadelphia: United, 1925), 459.

28. David Starr Jordan, quoted as representative in Harold Bulce, "Avatars of the Almighty," *Cosmopolitan Magazine* 47 (1909), 201. See also Marsden, *Fundamentalism and American Culture*, 130–31, 267–69.

29. Bryan, *In His Image*, 125. The "Menace of Darwinism" speech appeared as chapter 4 of this book, from which these quotes were taken.

30. William Jennings Bryan, *The Bible and Its Enemies* (Chicago: Bible Institute, 1921), 39.

31. Bryan, *In His Image*, 94.

32. Ibid., 98, 100.

33. Numbers, *The Creationists*, 43.

34. Bryan, *In His Image*, 93.

35. Ibid., 122.

36. William Jennings Bryan, quoted in Levine, *Defender of the Faith*, 277 (emphasis added).

37. William Bell Riley to William Jennings Bryan, 7 February 1923, in Bryan Papers.

38. "The Evolution Controversy," *Christian Fundamentals in Schools and Churches* 4 (April–June 1922), 5.

39. William Bell Riley, "Shall We Tolerate Longer the Teaching of Evolution?" *Christian Fundamentals in Schools and Churches* 5 (January–March 1923), 82.

40. W. B. Riley, "The Theory of Evolution Tested by Mathematics," in Trollinger, ed., *Anti-evolution Pamphlets*, 148.

41. William Vance Trollinger, Jr., "Introduction," in Trollinger, ed., *Anti-evolution Pamphlets*, xvii–xix.

42. Bryan, "Speech to Legislature," in William Jennings Bryan, *Orthodox Christianity Versus Modernism* (New York: Revell, 1923), 46.

43. Bryan, *In His Image*, 243. See also William Jennings Bryan, "Applied Christianity," *The Commoner*, May 1919, 11.

44. William Bell Riley to Charles S. Thomas, 1 July 1925, in Bryan Papers.

45. William Jennings Bryan, *Is the Bible True?* (Nashville: private printing, 1923), 15.

46. For estimates of popular support by current scholars, see Levine, *Defender of the Faith*, 270–71; Numbers, *The Creationists*, 44–45.

47. William Jennings Bryan, quoted in Levine, *Defender of the Faith*, 218.

48. Bryan, *In His Image*, 122.

49. Bryan, *Seven Questions in Dispute*, 154.

50. Bryan, "God and Evolution," sec. 7, p. 11.

51. Bryan, "Speech to Legislature," 48.

52. William Jennings Bryan, "Prohibition," *The Outlook* 133 (1923), 263.

53. Bryan, "Speech to Legislature," 45–46.

54. Edger Lee Masters, "The Christian Statesman," *The American Mercury* 3 (1924), 391.

55. William Jennings Bryan, quoted in "Progress of Anti-Evolution," *Christian Fundamentalist* 2 (1929), 13.

56. Bryan and Bryan, *Memoirs*, 179–80.

57. "A Remarkable Man," *Commercial Appeal* (Memphis), 29 April 1925, p. 6.

58. "Are People People?" *Chicago Tribune*, 20 June 1923, p. 8.

59. Amendment to 1923 Okla. House Bill 197.

60. 1923 Fla. House Concurrent Resolution 7.

61. William Jennings Bryan, "W. G. N. Put 'On Carpet,' Gets a Bryan Lashing," *Chicago Tribune*, 20 June 1923, p. 14 (contains quote and affirmed that "my views are set forth in" the Florida resolution).

62. Bryan, *In His Image*, 103–4 (emphasis in original).

63. William Jennings Bryan to Florida State Senator W. J. Singleton, 11 April 1923, in Bryan Papers.

64. "Memphis This Week Is Baptist Citadel," *Commercial Appeal* (Memphis), 11 May 1925, p. 1.

65. Edwin Conklin, "The Churches," in Philip M. Hamer, ed., *Tennessee—A History, 1672–1932*, vol. 2 (New York: American Historical Society, 1933), 826–27.

66. T. H. Alexander, "Biography," in *Austin Peay: A Collection of State Papers and Public Addresses* (Kingsport, Tenn.: Southern, 1929), xxx (quote); Joseph H. Parks and Stanley J. Folmsbee, *The Story of Tennessee* (Norman, Okla.: Harlow, 1963), 374; Billy Stair, "Religion, Politics, and the Myth of Tennessee Education," *Tennessee Teacher* 45 (1978), 19–20.

67. Austin Peay, "Address to Graduation Class of Carson and Newman College," in *Austin Peay*, 433–34.

68. Bryan, *Is the Bible True?* 3.

69. "Bryan's Latest," *Nashville Banner*, 28 January 1925, p. 8.

70. "Bryan Angrily Denies He's a Millionaire," *Commercial Appeal* (Memphis), 28 April 1925 , p. 1; "Remarkable Man," p. 6.

71. "Legislature Begins Drive on Evolution," *Commercial Appeal* (Memphis), 21 January 1925, p. 1.

72. *Journal of the House of Representatives of Tennessee* (1925 Reg. Sess.), 180.

73. John W. Butler, quoted in "Dayton's 'Amazing' Trial," *The Literary Digest* 86 (25 July 1925), 7.

74. 1925 Tenn. House Bill 185.

75. Bryan, *Is the Bible True?* 15–16.

76. This summary of House action was compiled from *Journal of the House*, 248; "Evolution In Schools Barred by the House," *Commercial Appeal* (Memphis), 28 January 1925, p. 1; "Peay Master of Assembly on Tax Plans," *Chattanooga Times*, 28 January 1925, p. 1; "Peay Program Is Voted by Solons," *Knoxville Journal*, 28 January 1925, p. 12; Ralph Perry, "Bar Teaching of Evolution," *Nashville Banner*, 28 January 1925, p. 3.

77. E. M. Matthews to Editor, *Nashville Banner*, 31 January 1925, p. 4.

78. Lee Wilkerson to Editor, *Nashville Banner*, 29 January 1925, p. 8.

79. Dillon J. Spottswood to Editor, *Nashville Tennessean*, 4 February 1925, p. 4.

80. Atha Hardy to Editor, *Nashville Tennessean*, 4 February 1925, p. 4.

81. Thomas Page Gore to Editor, *Nashville Banner*, 27 February 1925, p. 6.

82. "And Others Call It God," *Nashville Tennessean*, 1 February 1925, p. 4.

83. "Monkey Business," reprinted in "State Press Comment," *Knoxville Journal*, 11 February 1925, p. 6.

84. Louisville Courier-Journal, "Darwinism Done For," *Chattanooga Times*, 1 February 1925, p. 16.

85. "Lie Is Passed in Legislature," *Nashville Banner*, 5 February 1925, p. 7.

86. "Proceedings in Legislature," *Nashville Tennessean*, 11 March 1925, p. 8.

87. "Bill on Evolution Draws Fire Pastor," *Chattanooga Times*, 9 February 1925, p. 7.

88. "Baptists for Science in Church Colleges," *Commercial Appeal* (Memphis), 5 February 1925, p. 5.

89. Quoted in Kenneth M. Bailey, "The Enactment of Tennessee's Anti-Evolution Law," *Journal of Southern History* 41 (1950), 477.
90. *Journal of the Senate of Tennessee* (1925 Reg. Sess.), 214, 254, 286, 352.
91. Sam Edwards to Editor, *Nashville Banner*, 4 February 1925, p. 8.
92. J. R. Clerk to Editor, *Nashville Tennessean*, 5 February 1925, p. 4.
93. J. W. C. Church to Editor, *Nashville Tennessean*, 8 February 1925, p. 4.
94. Mrs. E. P. Blair to Editor, *Nashville Tennessean*, 16 March 1925, p. 4.
95. Dan Goodman to Editor, *Nashville Tennessean*, 6 February 1925, p. 4.
96. John A. Shelton to William Jennings Bryan, 5 February 1925, in Bryan Papers.
97. William Jennings Bryan to John A. Shelton, 9 February 1925, in Bryan Papers.
98. This review and the following summary of Sunday's crusade was compiled from daily articles in the 6 to 23 February 1925 issues of the Memphis *Commercial Appeal*, which included both daily news reports and a complete transcript of each sermon.
99. Ibid.
100. "First Verse of Bible Key to All Scripture," *Commercial Appeal* (Memphis), 13 March 1925, p. 11.
101. This summary of Senate action was compiled from "Evolution Is Given Hard Jolt," *Knoxville Journal*, 14 March 1925, p. 1; Howard Eskridge, "Senate Passes Evolution Bill," *Nashville Banner*, 13 March 1925, p. 1; "Legislators Bar Teaching Evolution," *Chattanooga Times*, 14 March 1925, p. 2; "Proceedings," p. 8; *Journal of the Senate of Tennessee* (1925 Reg. Sess.), 516–17; Thomas Fauntleroy, "Darwinism Outlawed in Tennessee Senate," *Commercial Appeal* (Memphis), 14 March 1925, p. 1.
102. Eskridge, "Senate Passes Evolution Bill," 1.
103. Excerpts from these letters to the governor appeared in "Peay Opens His Ape Law Letters," *Nashville Banner*, in Peay Papers, GP 40–13 (including quotation about Middle Ages); James L. Graham to Austin Peay, 18 March 1925, in Peay Papers, GP 40–13; James W. Mayor to Austin Peay, 14 March 1925, in Peay Papers, GP 40–13.
104. H. A. Morgan to Austin Peay, 9 February 1925, in Peay Papers, GP 40–24; "Anti-Evolution Bill Stirs Tennessee," *Atlanta Journal*, 24 May 1925, p. 11.
105. W. M. Wood to Austin Peay, 14 March 1925, in Peay Papers, GP 40–13.
106. Austin Peay, "Message from the Governor," 23 March 1925, in *Journal of the House of Representatives of Tennessee* (1925 Reg. Sess.), 741–45 (emphasis added). In his classic account of the Scopes trial, *Six Days of Forever? Tennessee v. John Thomas Scopes* (London: Oxford University Press, 1958), Ray Ginger uses this and other evidence to argue that the Tennessee antievolution statute was a symbolic protest rather than a serious law. It is clear to me, however, that antievolutionists took this statute seriously—they expected it to compel compliance. Bryan simply thought that, owing to the law-abiding nature of school-

teachers, the law would be self-enforcing rather than require active monitoring. What he did not anticipate was that some respected citizens would protest the statute by intentionally flaunting it. If there was a symbolic protest here, it was by opponents of the law in disobeying it rather than by proponents of the law in enacting it.

107. "Give Up Schools Before Bible Is Peay's Attitude," *Nashville Tennessean*, 27 June 1925, p. 1; Peay, "Message," 745.

108. W. J. Bryan to Austin Peay, undated telegram, in Bryan Papers.

109. Peay, "Message," 743, 745; Bryan to Shelton, 9 February 1925.

Chapter Three. In Defense of Individual Liberty

1. James Harvey Robinson, *The Mind in the Making: The Relation of Intellect to Social Reform* (New York: Harper, 1921), 181–86.

2. Woodrow Wilson, "War Message, April 2, 1917," in *Papers of Woodrow Wilson*, vol. 41 (Princeton, N.J.: Princeton University Press, 1983), 519–27.

3. Postmaster General Albert S. Burleson, quoted in Paul L. Murphy, *World War I and the Origin of Civil Liberties in the United States* (New York: Norton, 1979), 98.

4. Roger N. Baldwin to W. D. Collins, 25 January 1918, in ACLU Archives, vol. 26.

5. Norman Thomas, "War's Heretics: A Plea for the Conscientious Objector," reprinted in *The Survey* 33 (1917), 391–94 (quote at 394).

6. Roger N. Baldwin, quoted in Samuel Walker, *In Defense of American Liberties: A History of the ACLU* (New York: Oxford University Press, 1990), 39.

7. Walker, *In Defense of American Liberties*, 21.

8. Robinson, *Mind in the Making*, 180–82.

9. Ibid., 186.

10. Ibid., 187–88.

11. Quoted in "The Real Motives Back of the Tennessee Evolution Case," *National Bulletin* (Military Order of the World War), June 1925, p. 3.

12. Roger N. Baldwin, quoted in Peggy Lamson, *Roger Baldwin: Founder of the American Civil Liberties Union* (Boston: Houghton Mifflin, 1976), 124.

13. Roger N. Baldwin, quoted in Walker, *In Defense of American Liberties*, 46–47.

14. Gitlow v. New York, 268 U.S. 652, 666 (1925).

15. Schenck v. United States, 249 U.S. 47, 51–52 (1919).

16. Oliver Wendell Holmes, quoted in Gerald Gunter, *Learned Hand: The Man and the Judge* (New York: Knopf, 1994), 163.

17. Abrams v. United States, 250 U.S. 616, 828–30 (Holmes, J., dissenting, 1919) (emphasis added).

18. American Civil Liberties Union, *The Fight for Free Speech* (New York: American Civil Liberties Union, 1921), 6–8.

19. Roger N. Baldwin, in Walker, *In Defense of American Liberties*, 52.

20. Arthur Garfield Hays, *City Lawyer: The Autobiography of a Law Practice* (New York: Simon and Schuster, 1942), 227. Despite this position on free speech, Hays regularly brought libel actions on behalf of his clients and himself.

21. Arthur Garfield Hays, *Let Freedom Ring* (New York: Liveright, 1928), xi.

22. Ibid., xx.

23. Walker, *In Defense of American Liberties*, 53.

24. Arthur Garfield Hays, *City Lawyer*, 227.

25. David E. Lilienthal, "Clarence Darrow," *Nation* 124 (1927), 417.

26. In a characteristic comment on the topic, Darrow observed that "the only thing I ever saw that seemed to have free will was an electric pump I had once on a summer vacation. Every time we wanted it to go, it stopped. I couldn't think of anything except free will, and all of a sudden when we knew nothing about it, it started again." In Clarence Darrow and Will Durant, *Is Man a Machine?* (New York: League for Public Discussion, 1927), 51.

27. Will Herberg, *Protestant, Catholic, Jew* (Garden City: Doubleday, 1960), 259–60.

28. Kevin Tierney, *Darrow: A Biography* (New York: Croswell, 1979), 85.

29. Robert G. Ingersoll, "Reply to Dr. Lymann Abbott," in Robert G. Ingersoll, *The Works of Robert G. Ingersoll*, vol. 4 (New York: Dresden, 1903), 463.

30. Clarence Darrow, *The Story of My Life* (New York: Grosset, 1932), 409; Clarence Darrow, "Why I Am an Agnostic," in Clarence Darrow, *Verdicts Out of Court*, Arthur Weinberg and Lila Weinberg, eds. (Chicago: Quadrangle, 1963), 434.

31. Clarence Darrow, quoted in Lilienthal, "Darrow," 419.

32. Darrow, "Why I Am an Agnostic," 436.

33. For example, Darrow, *My Life*, 382–423 (quotes from 383 and 419).

34. Darrow, *My Life*, 408–13.

35. "Darrow Asks W. J. Bryan to Answer These," *Chicago Daily Tribune*, 4 July 1923, p. 1.

36. See W. B. Norton, "Bryan's Ailment Is Intolerance, Pastor's Assert," *Chicago Daily Tribune*, 28 June 1923, p. 3.

37. Darrow, *My Life*, 249.

38. John Haynes Holmes, *I Speak for Myself: The Autobiography of John Haynes Holmes* (New York: Harper, 1959), 263. The philosopher Will Durant made a similar observation during a debate with Darrow. Darrow and Durant, *Is Man a Machine?* 45.

39. Hays, *City Lawyer*, 221.

40. For example, Henry R. Linville, *The Biology of Man and Other Organisms* (New York: Harcourt Brace, 1923), 4–5. See also Henry R. Linville and Henry A. Kelly, *A Text-book in General Zoology* (Boston: Ginn, 1906).

41. American Civil Liberties Union, *The Fight for Free Speech* (New York: American Civil Liberties Union, 1921), 17–18.

42. Lusk Committee, quoted in Robinson, *Mind in the Making*, 190–91.

43. Walker, *In Defense of American Liberties*, 59. The ACLU sought to get its message into public schools during the twenties both by providing schools speakers and helping high school debaters prepare for debates on free-speech issues. See, e.g., Roger N. Baldwin to College and High School Debating Societies, 17 October 1924, ACLU Archives, vol. 253. A typical example of classroom "Americanism" materials that surfaced in many places was the U.S. Bureau of Education's 1923 American Education Week curriculum, which lauded the country's founding fathers and military exploits.

44. Darrow, *My Life*, 25.

45. Georgia Supreme Court, quoted in William Seagle, "A Christian Country," *American Mercury* 6 (1925), 233.

46. 1915 Tenn. Acts, ch. 102.

47. Joseph Story, *Commentaries on the Constitution of the United States*, vol. 2 (Boston: Little & Brown, 1851), 590–97.

48. Andrew Dickson White describes the episode at length in Andrew Dickson White, *A History of the Warfare of Science with Theology in Christendom*, vol. 1 (London: Macmillan, 1896), 313–16. See also Paul K. Conkin, *Gone With the Ivy: A Biography of Vanderbilt University* (Knoxville: University of Tennessee Press, 1985), 50, 60–62.

49. George M. Marsden, *The Soul of the American University: From Protestant Establishment to Established Non-Belief* (New York: Oxford University Press, 1994), 130.

50. White, *History of the Warfare of Science*, vol. 1, 315.

51. "The Case of Professor Mecklin," *Journal of Philosophical, Psychological, and Scientific Methods* 11 (1918), 67–81.

52. Arthur O. Lovejoy, "Organization of the American Association of University Professors," *Science* 41 (1915), 152.

53. "General Report of the Committee on Academic Freedom and Academic Tenure," *Bulletin of the American Association of University Professors* 1 (December 1915), 21, 23, 27, 29–30.

54. "Report on the University of Tennessee," *Bulletin of the American Association of University Professors* 10 (1924), 217.

55. Ibid., 213–59 (quotes at 217 and 255). See also Jonas Riley Montgomery, Stanley J. Folmebee, and Lee Seifern Greene, *To Foster Knowledge: A History of the University of Tennessee, 1794–1970* (Knoxville: University of Tennessee Press, 1984), 185–87.

56. "Report on Tennessee," 56–63 (quotes); Montgomery, Folmebee, and Greene, *To Foster Knowledge*, 187–89.

57. Joseph V. Dennis, "Presidential Address," *Bulletin of the American Association of University Professors* 10 (1924), 26–28.

58. "Report of Committee M," *Bulletin of the American Association of University Professors* 11 (1925), 93–94.

59. Henry R. Linville, "Tentative Statement of a Plan for Investigating Work on Free-Speech Cases in Schools and Colleges," in ACLU Archives, vol. 248.

60. Harry F. Ward, "MEMORANDUM on Academic Freedom," ACLU Archives, vol. 248.

61. Harry F. Ward and Henry R. Linville, "Freedom of Speech in Schools and Colleges: A Statement by the American Civil Liberties Union, June, 1924," ACLU Archives, vol. 248.

62. "Free Speech in Colleges Tackled by New Group—Civil Liberties Union Forms Committee to Act in Cases of Interference with Students and Teachers," 22 October 1924, Press Release, ACLU Archives, vol. 248. See also John Haines Holmes and Roger Baldwin to "Colleges," 15 November 1924, ACLU Archives, vol. 248.

63. Lucille Milner, *Education of an American Liberal: An Autobiography of Lucille Milner* (New York: Horizon, 1954), 161–62.

64. Roger N. Baldwin, "Dayton's First Issue," in Jerry R. Tomkins, ed., *D-Days at Dayton: Reflections of the Scopes Trial* (Baton Rouge: Louisiana State University Press, 1965), 56.

65. "Cries at Restrictive Laws," *New York Times*, 26 April 1925, in ACLU Archives, vol. 273.

66. "Plan Assault on State Law on Evolution," *Chattanooga Daily Times*, 4 May 1925, p. 5; "Anti-Evolution Law Won't Affect Elementary Schools," *Jackson Sun*, 29 March 1925, in Peay Papers, GP 40–13.

Chapter Four. Choosing Sides

1. *Why Dayton, of All Places?* (Chattanooga: Andrews, 1925), 3.

2. The Census Bureau did not list Dayton in the 1870 census and placed its population at only 200 persons in 1880. The 1890 census reported a population of 2,719 for Dayton in 1890, but by the 1900 census the figure had dropped to 2,004. Thereafter, the Census Bureau ceased to separately list towns under 2,500 and Dayton dropped off the list. Boosted by agricultural development, however, the Rhea County population continued to grow. See U.S. Census Bureau, *1880 Census: Population*, vol. 1 (Washington, D.C.: Government Printing Office, 1883), 338; U.S. Census Bureau, *1900 Census: Population*, vol. 1, pt. 1 (Washington, D.C.: Government Printing Office, 1901), 373.

3. "Rappleyea Rapped," *Chattanooga Times*, 19 May 1925, p. 5. Some secondary sources spelled this surname "Rappelyea," but contemporary sources used "Rappleyea."

4. "Was Converted Through Science," *Chattanooga Times*, 21 May 1925, p. 2; G. W. Rappleyea to Editor, *Chattanooga Times*, 19 May 1925, p. 5. When his devout Roman Catholic mother later read about his pivotal role in these events, she chided him, "You always had lots of book sense, but never any common sense." "Was Converted Through Science," p. 2.

5. Fred E. Robinson, in Warren Allem, "Backgrounds of the Scopes Trial at Dayton, Tennessee," Master's thesis, University of Tennessee, 1959, p. 58.

6. Untitled typed statement of S. K. Hicks replying to charges before the Tennessee state bar association, in Hicks Papers.

7. John T. Scopes, *The Center of the Storm: Memoirs of John T. Scopes* (New York: Holt, 1967), 58–59; Juanita Glenn, "Judge Still Recalls 'Monkey Trial'—50 Years Later," *Knoxville Journal*, 11 July 1975, p. 17; Allem, "Backgrounds of the Scopes Trial," 60–61.

8. T. W. Callaway, "Father of Scopes Renounced Church," *Chattanooga Times*, 10 July 1925, p. 1.

9. Arthur Garfield Hays, *Let Freedom Ring* (New York: Liveright, 1928), 33.

10. Sue K. Hicks in Glenn, "Judge Still Recalls," 17. Other firsthand versions reported Scopes saying, "I'll be willing to stand trial," or "I don't care. Go ahead." Indeed, Scopes recalled it both ways himself at different times and said that he was given two opportunities to "back down." John T. Scopes, in "Chance Conversation Started Scopes Case," *Knoxville Journal*, 30 May 1925, p. 1; Scopes, *Center of the Storm*, 60. See also Sue K. Hicks quoting Scopes in Allem, "Backgrounds of the Scopes Trial," 60.

11. Walter White, in Allem, "Backgrounds of the Scopes Trial," 61.

12. "Arrest Under Evolution Law," *Nashville Banner*, 6 May 1925, p. 1.

13. "Cheap Publicity," *Nashville Tennessean*, 23 June 1925, p. 4.

14. "Darwin Bootlegger Arrested by Deputy," *Commercial Appeal* (Memphis), 7 May 1925, p. 1.

15. John P. Fort, "Final Resolution Demands Chattanooga Cease Move to Bring New Case," *Chattanooga News*, 19 May 1925; "One Evolutionist Out of Hundred," *Chattanooga Times*, 11 July 1925, p. 1; H. L. Mencken, "The Monkey Trial: A Reporter's Account," in Jerry R. Tompkins, ed., *D-Days at Dayton: Reflections on the Scopes Trial* (Baton Rouge: Louisiana State University Press, 1965), 44 (reprint of 15 July 1925 article).

16. The 1920 federal census reported that only 6.5 percent of the Rhea County population was "Negro," precisely one-third the Tennessee state percentage. U.S. Census Bureau, *1920 Census: Population*, vol. 3 (Washington, D.C.: Government Printing Office, 1922), 961, 167.

17. Mencken, "Monkey Trial," 36–37 (reprint of 9 July 1925 article).

18. "Rebuke to the 'Antis'," *Chattanooga Times*, 4 June 1925, p. 4; Edwin Mims, "Address to Southern Conference on Education," in Mims Papers; "Peay Not to Visit Dayton for Trial," *Nashville Banner*, 16 June 1925, p. 1; "Doubts Legality of Special Term," *Chattanooga Times*, 24 May 1925, p. 9; "J. Will Taylor's Comments," *Nashville Banner*, 26 May 1925, p. 1.

19. "Cheap Publicity," p. 4; "Hungered and Thirsted for Publicity," *Knoxville Journal*, 18 July 1925, p. 6; "A Humiliating Proceeding," *Chattanooga Times*, 8 July 1925, p. 4; "Dayton Now Famous," *Nashville Banner*, 26 May 1925, p. 8; "Tennessee's Opportunity," *Nashville Banner*, 12 July 1925, p. 1.

20. "The Dayton Serio-Comedy," *Chattanooga Times*, 24 June 1925, p. 4.

21. "Southerners Open the Exposition," *New York Times*, 12 May 1925, p. 11.

22. "The South and Its Critics," *Chattanooga Times*, 8 May 1925, p. 1.

23. "Come South," *Nashville Banner*, 6 May 1925, p. 8, and "Arrest Under Law," p. 1.

24. Scopes, *Center of the Storm*, 63.

25. "Scopes Held for Trial Under Evolution Law," *Commercial Appeal* (Memphis), 10 May 1925, p. 1; "Scopes Held to Grand Jury in Evolution Test," *Nashville Tennessean*, 10 May 1925, p. 1; Scopes, *Center of the Storm*, 63–65.

26. "Evolution Taught at Central High," *Chattanooga Times*, 19 May 1925, p. 5.

27. "Dayton to Raise Advertising Fund," *Chattanooga Times*, 23 May 1925, p. 15; "Dayton Seeks Pup Tents and Loud Speakers for Scopes Trial Crowd," *Nashville Tennessean*, 23 May 1925, p. 1; "Set Stage for Evolution Case," *Nashville Banner*, 24 May 1925, p. 1; "Preparations Begin for Evolution Trial," *Knoxville Journal*, 6 June 1925, p. 1; "Nation Divided on Darwinism as Trial Looms," *Nashville Tennessean*, 29 May 1925, p. 1.

28. "Trial Can Be Held in June, Says Judge," *Chattanooga Times*, 21 May 1925, p. 2; "H. G. Wells May Fight Bryan in Scopes Case," *Commercial Appeal* (Memphis), 15 May 1925, p. 35.

29. "You May Not Be for Him, but, Nevertheless, There He Is," *Columbus Dispatch*, 14 July 1925, p. 4; W. J. Bryan to Cartoonist, *Columbus Dispatch*, 27 July 1925, p. 1.

30. "Material Criticism Decries Supernatural," *Commercial Appeal* (Memphis), 5 May 1925, p. 1.

31. "Commoner Believes Evolution Tommyrot," *Commercial Appeal* (Memphis), 11 May 1925, p. 1.

32. Ibid., p. 1; "Radical Enemies of Evolution Forced to Acknowledge Defeat," *Commercial Appeal* (Memphis), 15 May 1925, p. 1; W. B. Riley, "The World's Christian Fundamentals Association and the Scopes Trial," *Christian Fundamentals in School and Church* 7 (October–December 1925), 37; "Bryan May Be in Case," *Nashville Banner*, 12 May 1925, p. 1.

33. The Memphis Press to Rhea County, 14 May 1925, in Hicks Papers.

34. Sue K. Hicks to the *Memphis Press*, 14 May 1925, in Hicks Papers; Sue K. Hicks to William J. Bryan, 14 May 1925, reprinted in William Jennings Bryan and Mary Baird Bryan, *The Memoirs of William Jennings Bryan* (Philadelphia: United, 1925), 483.

35. W. J. Bryan to Sue Hicks, 20 May 1925, in Hicks Papers.

36. "Darrow Falls Back on Omar," *Commercial Appeal* (Memphis), 18 April 1925, p. 6.

37. Clarence Darrow, *The Story of My Life* (New York: Grosset, 1932), 249.

38. "Darrow Ready to Aid Prof. Scopes," *Nashville Banner*, 16 May 1925, p. 1. Bryan's daughter Grace later blamed the participation of Malone in the Scopes defense on his desire to "'get even' because of the severe rebuke father gave him which resulted in his dismissal from the State Department" when Bryan served as Secretary of State. Grace Dexter Bryan to Sue K. Hicks, 12 April 1940, in Hicks Papers.

39. "Make It Bryan vs. Darrow," *St. Louis Post-Dispatch*, 14 May 1925, p. 20.

40. Forrest Bailey to Walter Lippmann, 12 June 1926, in ACLU Archives, vol. 311.

41. American Civil Liberties Union, Minutes of 6/8/25 Executive Committee Meeting, in ACLU Archives, vol. 279.

42. Scopes, *Center of the Storm*, 70–72.

43. Bailey to Lippmann, 12 June 1926.

44. W. H. Pitkin to Felix Frankfurter, 10 November 1926, in ACLU Archives, vol. 299.

45. "Scopes Dined, Says Fight Is for Liberty," *New York Times*, 11 June 1925, p. 1; "Malone Will Not Be Goat," *Nashville Banner*, 11 June 1925, p. 1; "Jazz Faction Puts Malone Back in Case," *Chattanooga Times*, 11 June 1925, p. 1.

46. "Darrow Likens Bryan to Nero," *Nashville Banner*, 18 May 1925, p. 1.

47. For example, Brewer Eddy to John R. Neal, 10 September 1925, in ACLU Archives, vol. 274.

48. Bryant Harbert, "Darrow an Atheist, Is Bryan's Answer," *Commercial Appeal* (Memphis), 23 May 1925, p. 1.

49. "Bryan Hissed and Cheered in Evolution Speech," *Nashville Tennessean*, 19 May 1925, p. 1; William Jennings Bryan to James W. Freedman, 10 June 1925, in Bryan Papers. For the *New York Times* editorial position, see "The End Is in Sight at Dayton," *New York Times*, 18 July 1925, p. 12; "Ended at Last," *New York Times*, 22 July 1925, p. 18.

50. "Dayton Jolly as Evolution Trial Looms," *Chattanooga Times*, 21 May 1925, p. 1; "Dayton to Raise Advertising Fund," *Chattanooga Times*, 23 May 1925, p. 15.

51. "Clarence Darrow Retires," *Commercial Appeal* (Memphis), 27 April 1925, p. 6.

52. "Scopes' Legal Advisors Split on Outside Aid," *Chattanooga Times*, 29 May 1925, p. 1; "Dayton Jolly," 1; "Darrow–Malone Defense Scopes Riles Dayton," *Knoxville Journal*, 28 May 1925, p. 1; "Evolution Trial Raises Two Sharp Issues," *New York Times*, 31 June 1925, sec. 9, p. 4; "The Scopes Defense," *Commercial Appeal* (Memphis), 30 May 1925, p. 8; "Scopes Glad to Have Help of Notables," *Chattanooga Times*, 30 May 1925, p. 1.

53. "Evolution Case Won't Test Truth of Theory, Says Neal," *Nashville Tennessean*, 16 May 1925, p. 1.

54. "Dr. Neal Swamped by Mail Over Scopes Case," *Knoxville Journal*, 28 May 1925, p. 1.

55. "Two Extreme Views," *Chattanooga Times*, 1 July 1925, p. 4.

56. "Not in Favor of Extra Term of Rhea Court," *Chattanooga Times*, 21 May 1925, p. 2.

57. "Says Evolution Law Wholesome Statute," *Chattanooga Times*, 24 May 1925, p. 9.

58. Philip Kinsley, "Scopes Indicted for Teaching Evolution," *Commercial Appeal* (Memphis), 26 May 1925, p. 1.

59. "Prompt Action by Grand Jury," *Nashville Banner*, 25 May 1925, p. 1; Kinsley, "Scopes Indicted," p. 1; "Scopes Is Indicted in Tennessee for Teaching Evolution," *New York Times*, 26 May 1925, p. 1; "Jury Foreman in Scopes Case Is Evolutionist," *Evening World* (New York), 26 May 1925, p. 1.

60. "Judge's Own Views," *Nashville Banner*, 25 May 1925, p. 5; Kinsley, "Scopes Indicted," 2.

61. "Trial July 10 Suits Bryan," *Nashville Banner*, 26 May 1925, p. 1.

62. Kinsley, "Scopes Indicted," 1.

Chapter Five. Jockeying for Position

1. "Butler Denounces New Barbarians," *New York Times*, 4 June 1925, p. 3.

2. "Antievolution Law Termed Outrageous," *Nashville Banner*, 28 June 1925, p. 1; "Tennessee Hit by Dr. Potter," *Nashville Banner*, 15 June 1925, p. 1.

3. "Shaw and Coleman on Scopes Trial," *New Leader* (New York) 25 July 1925, p. 6 (reprint of Shaw's pretrial speech); David M. Church, "Net of Dayton Trial Spreads," *Nashville Banner*, 7 June 1925, sec. 2, p. 7; Frederick Kuh, "Ape Case Loosens Up Tongue of Einstein," *Pittsburgh Sun*, 22 June 1925, p. 10.

4. "Scopes Dined Says Fight Is for Liberty," *New York Times*, 11 June 1925, p. 1.

5. Oliver H. P. Garett, "Colby Enters Scopes Case: Darrow Chief," *Chattanooga Times*, 10 June 1925, p. 1 (reprint of *New York World* article).

6. "Scopes Here Grins, Does Not Know If He Is a Christian," *New York World*, 7 June 1925, p. 1; Edward Levinson, "Man and Monkey: An Interview with Scopes," *New Leader* (New York), 11 July 1925.

7. "Scopes Rests Hope in U.S. Constitution and Supreme Court," *Washington Post*, 13 June 1925, p. 1.

8. Henry Fairfield Osborn, *Evolution and Religion in Education* (New York: Scribner's, 1925), 34, 90, 96, 117, 122 (reprint of Osborn's 1925 *The Earth Speaks to Bryan*, with added chapters); "Science and Showmanship," *New York World*, 14 July 1925, p. 10.

9. "Dr. Osborn Advises Scopes on Defense," *New York Times*, 9 June 1925, p. 5.

10. "Scientists Pledge Support to Tennessee Professor Arrested for Teaching Evolution," *Daily Science News Bulletin*, 18 May 1925 (press release), in ACLU Archives, vol. 273.

11. George Hunter, *A Civic Biology* (New York: American, 1914), 261.

12. C. B. Davenport, "Evidences for Evolution," *Nashville Banner*, 1 June 1925, p. 6.

13. Henry Fairfield Osborn, "Osborn States the Case for Human Evolution," *New York Times*, 12 July 1925, sec. 8, p. 1; Kenneth Kyle Bailey, "The Anti-Evolution Crusade of the Nineteen-Twenties," Ph.D. diss., Vanderbilt University, 1953, p. 127; Luther Burbank to John Haynes Holmes, 29 July 1925,

in ACLU Archives, vol. 274; Ray Ginger, *Six Days or Forever? Tennessee versus John Thomas Scopes* (Boston: Beacon, 1958), p. 79.

14. "'Monkey Law' in Limelight," *Nashville Banner*, 22 May 1925, p. 1.

15. "Declares Bryan Befogs the Issue," *New York Times*, 18 May 1925, p. 8.

16. "Says Scopes Trial Will Help Religion," *New York Times*, 15 June 1925, p. 18; "Bible and Evolution Conflict, Says Potter," *Chattanooga Times*, 2 July 1925, p. 2.

17. H. L. Mencken, "The Monkey Trial: A Reporter's Account," in Jerry R. Tomkins, ed., *D-Days at Dayton: Reflections of the Scopes Trial* (Baton Rouge: Louisiana State University Press, 1965), 40 (reprint of 13 July 1925 article).

18. Transcript, 75.

19. "Dean Divinity School Thinks Bible Is Not in Conflict with Evolution," *Chattanooga Times*, 10 July 1925, p. 14; Transcript, 224–25.

20. "Sees Bryan as a Pharisee," *New York Times*, 18 May 1925, p. 8; "Two Extreme Views," *Chattanooga Times*, 1 July 1925, p. 4; "Tennessee Held Up to Scorn by Aked," *Nashville Banner*, 12 June 1925, p. 2.

21. "Evolution Is Discussed by Two Ministers in Knoxville," *Knoxville Journal*, 8 June 1925, p. 5.

22. "V. U. Seniors Hear Theologian Rank Darwin as Saint," *Nashville Tennessean*, 8 June 1925, p. 1.

23. Transcript, 223–24.

24. Transcript, 229.

25. William Jennings Bryan, *Seven Questions in Dispute* (New York: Revell, 1924), 128.

26. "Real Religion and Real Science," *Commercial Appeal* (Memphis), 26 July 1925, sec. 1, p. 4.

27. "Tennessee Hit by Dr. Potter," p. 1; Herbert Sanborn, "Four Species of Evolution," *Nashville Banner*, 5 July 1925, p. 2.

28. "Deny Science Wars Against Religion," *New York Times*, 25 May 1923, p. 1.

29. "Evolution Given Hard Jolt," *Knoxville Journal*, 14 March 1925, p. 1.

30. "Resolution Aimed at Tennessee Law," *Nashville Banner*, 1 July 1925, p. 21; "Educators Taboo Evolution Question," *Nashville Banner*, 1 July 1925, p. 21.

31. "Pastor Compares Darrow, Devil," *Knoxville Journal*, 1 July 1925, p. 2.

32. J. Frank Norris to W. J. Bryan, n.d. [June 1925], in Bryan Papers.

33. "Billy Sunday Not to Go to Dayton," *Nashville Banner*, 7 July 1925, p. 9.

34. Ronald L. Numbers, "Introduction," in Ronald L. Numbers, ed., *Creation–Evolution Debates* (New York: Garland, 1995), ix.

35. W. J. Bryan to John Straton, 1 July 1925, in Bryan Papers.

36. "S. F. Debate on Evolution Ends in 'Tie'," *San Francisco Examiner*, 15 June 1925, p. 13.

37. "The San Francisco Debates on Evolution," in Numbers, ed., *Creation–Evolution Debates*, 196, 289–90, 364.

38. George McCready Price to W. J. Bryan, 1 July 1925, in Bryan Papers.

39. W. J. Bryan to Dorothy MacIver James, 9 May 1925, in Bryan Papers.

40. "Daily Editorial Digest," *Nashville Banner*, 22 May 1925, p. 8; "Weird Adventures of 200 Reporters At Tennessee Evolution Trial," *Editor & Publisher*, 18 July 1925, p. 1.

41. Edward Caudill, "The Roots of Bias: An Empiricist Press and Coverage of the Scopes Trial," *Journalism Monographs* 114 (July 1989), 32.

42. "Novel View in Evolution Row," *Chattanooga Times*, 6 July 1925, p. 12.

43. T. W. Callaway, "One Evolutionist Out of Hundred," *Chattanooga Times*, 11 July 1925, p. 1.

44. Ira Hicks to Sue K. Hicks, n.d. [mid June 1925], in Hicks Papers.

45. "Baptist Editor Supports Bryan," *Nashville Banner*, 13 June 1925, p. 1.

46. "Evolution in the Public Schools," *The Present Truth*, 1 July 1925, p. 1.

47. "Dayton Keyed Up for Opening Today of Trial of Scopes," *New York Times*, 10 July 1925, p. 1.

48. Philip Kinsley, "Invoke Divine Guidance for Evolution Case," *Chicago Tribune*, 10 June 1925, p. 1.

49. "Anti-Evolution Leagues Form Over Country," *Chattanooga Times*, 2 July 1925, p. 1.

50. "Bryan Discusses Tennessee Case," *Nashville Banner*, 2 June 1925, p. 3.

51. "Bryan Calls Attention to Decision in Oregon," *Chattanooga Times*, 4 June 1925, p. 1.

52. "Modernist Fires Back at Commoner," *Nashville Banner*, 20 May 1925, p. 1.

53. "People Will Settle Question, Says Bryan," *Chattanooga Times*, 3 July 1925, p. 1.

54. "Bryan Gets the Jump on Defense Lawyers," *Commercial Appeal* (Memphis), 9 July 1925, p. 1.

55. W. J. Bryan to Sue K. Hicks, 28 May 1925, in Hicks Papers.

56. Sue K. Hicks to Ira E. Hicks, 8 June 1925, in Hicks Papers.

57. Sue K. Hicks to Reese V. Hicks, 8 June 1925, in Hicks Papers.

58. Bryan to Hicks (emphasis in original).

59. "Bryan Outlines Issues," p. 1.

60. Sue Hicks to Reese Hicks; Sue Hicks to Ira Hicks.

61. W. J. Bryan to Dr. Howard A. Kelly, 17 June 1925, in Bryan Papers.

62. Sue Hicks to Ira Hicks.

63. Ira E. Hicks to Sue K. Hicks, 5 June 1925, in Hicks Papers; Sue K. Hicks to W. J. Bryan, 8 June 1925, in Hicks Papers.

64. W. J. Bryan to George McCready Price, 1 June 1925, in Bryan Papers.

65. George McCready Price to W. J. Bryan, 1 July 1925, in Bryan Papers.

66. Howard A. Kelly to W. J. Bryan, 15 June 1925, in Bryan Papers.

67. W. J. Bryan to Howard A. Kelly, 22 June 1925, in Bryan Papers.

68. W. J. Bryan to S. K. Hicks, 10 June 1925, in Hicks Papers.

69. Ibid.

70. Samuel Untermyer to W. J. Bryan, 25 June 1925, in Bryan Papers.

71. Herbert E. Hicks and Sue K. Hicks to W. J. Bryan, 10 June 1925, in Hicks Papers; Herbert E. Hicks and Sue K. Hicks to W. J. Bryan, 13 June 1925, in Hicks Papers; W. J. Bryan to S. K. Hicks, 16 June 1926, in Hicks Papers; Ginger, *Six Days or Forever?* 74–78.

72. Herbert E. Hicks and Sue K. Hicks to John T. Raulston, 1 July 1925, in Hicks Papers.

73. "Give Up Schools Before Bible Is Peay's Attitude," *Nashville Tennessean*, 27 June 1925, p. 1.

74. "Dayton Trial Will Be Brief, Nashville Lawyer Predicts," *Nashville Banner*, 9 June 1925, p. 10.

75. W. J. Bryan to W. B. Marr, 15 June 1925, in Darrow Papers.

76. "Stewart Predicts Act Will Stand Acid Test," *Chattanooga Times*, 11 June 1925, p. 2.

77. "Scopes Counsel Expect Trial to Last Month," *Nashville Tennessean*, 23 June 1925, p. 1; Arthur Garfield Hays, "The Strategy of the Scopes Defense," *The Nation*, 5 August 1925, pp. 157–58.

78. "Witnesses for Defense at Dayton," *Nashville Banner*, 26 June 1925, p. 20.

79. Bryan, *Seven Questions in Dispute*, 154; "Bryan Discusses Evolution Case," *Nashville Banner*, 2 June 1925, p. 3; Transcript, 172.

80. "No Such Thing as Evolution, Bryan Declares," *Chattanooga Times*, 2 June 1925, p. 1.

81. W. J. Bryan to W. B. Riley, 27 March 1925, in Bryan Papers; Transcript, 230.

82. Mencken, "Monkey Trial," 40 (reprint of 13 July 1925 article).

83. John T. Scopes, *The Center of the Storm: Memoirs of John T. Scopes* (New York: Holt, 1967), 78.

84. Letter from Sue K. Hicks to W. J. Bryan, 23 June 1925, in Hicks Papers.

85. "Darrow Declares He Is Always Seeking Truth," *Knoxville Journal*, 23 June 1925, p. 1; "Glut of Laws Threat Against All Freedom," *Commercial Appeal* (Memphis), 24 June 1925, p. 1; "Scopes Dined Says Fight Is for Liberty," p. 1; "Drab Views Bared Before Capacity Crowd in Speech," *Knoxville Journal*, 24 June 1925, p. 1.

86. Arthur Garfield Hays, *City Lawyer* (New York: Simon & Schuster, 1942), 212.

87. Hicks to Bryan, 23 June 1925; "Drab Views Bared," p. 1.

88. James Gibson, "Evolution Stalks State Bar Meeting," *Commercial Appeal* (Memphis), 17 June 1925, p. 1; "Keebler's Attack on Legislation Is Not Sustained," *Knoxville Journal*, 26 June 1925, p. 1.

89. "$10,000 Defense Fund Is Asked for Scopes; To Be Spent on Case in Lower Courts," *New York Times*, 22 June 1925, p. 1.

90. "Anti-Evolution Act Invasion of Rights—Malone," *Knoxville Journal*, 28 June 1925, p. 1.

91. "Scopes Counsel, Here, Plans for Evolution Case," *Chicago Tribune*, 30 June 1925, p. 6.
92. "Scopes Trial Food for Thought, Colby," *Knoxville Journal*, 25 June 1925, p. 1.
93. "Rogers Sidesteps Evolution Trial," *Nashville Banner*, 31 May 1925, p. 1.
94. "Scopes Trial Call Test Free Speech," *Chattanooga Times*, 25 June 1925, p. 5; "Malone Pays His Respects to W. J. Bryan," *Chattanooga Times*, 26 June 1925, p. 5; "Malone Says Constitution Facing Test," *Chattanooga Times*, 27 June 1925, p. 5; "Says Legislature Can't Make Morals," *Nashville Banner*, 28 June 1925, p. 5.
95. W. B. Marr to Sue and Herbert Hicks, 10 June 1925, in Hicks Papers.
96. "Dr. Neal Says Renaissance Now Dawning," *Chattanooga Times*, 30 June 1925, p. 1; "Scopes Dined Says Fight Is for Liberty," p. 1.
97. "Scopes Attorney Fight Dayton Trial," *New York Times*, 4 July 1925, p. 2; "Federal Judge Refuses to Take Jurisdiction in Scopes Case; Trial Opens Friday in Dayton," *Chattanooga Times*, 7 July 1925, p. 1; Scopes, *Center of the Storm*, 78–82.
98. "Federal Judge Refuses," p. 1.
99. Mencken, "Monkey Trial," 44 (reprint of 15 July 1925 article).
100. "Broadcast of Scopes Trial Unprecedented," *Chicago Tribune*, 5 July 1925, sec. 7, p. 6; Philip Kinsley, "Dayton Raises Curtain Soon," *Nashville Banner*, 5 July 1925, p. 6.
101. Mencken, "Monkey Trial," 43 (reprint of 15 July 1925 article). Mason's activities in Tennessee were discussed in Joe Maxwell, "Building the Church (of God in Christ)," *Christianity Today*, 8 April 1996, p. 25.
102. "Dayton Keyed Up for Opening Today of Trial of Scopes," *New York Times*, 10 July 1925, p. 1.
103. T. W. Callaway, "Dayton Bootblack Gives Preacher His Definition of Fundamentalism," *Chattanooga Times*, 8 July 1925, p. 1.
104. "Dayton Cheers the Commoner," *Chattanooga Times*, 8 July 1925, p. 1.
105. "Bryan in Dayton, Calls Scopes Trial Duel to the Death," *New York Times*, 8 June 1925, p. 1.
106. "Visitors Come on Every Train," *Nashville Banner*, 9 July 1925, p. 3.
107. W. C. Cross, "Bryan, Noted Orator, in Favor at Dayton," *Knoxville Journal*, 10 July 1925, p. 1.
108. "Bryan Threatens National Campaign to Bar Evolution," *New York Times*, 8 July 1925, p. 1.
109. Philip Kinsley, "Bryan Gets Jump on Defense Lawyers," *Commercial Appeal*, 9 July 1925, p. 1.
110. Transcript, 159.
111. "Malone Glad Trial Starts on a Friday," *Chattanooga Times*, 10 July 1925, p. 2.
112. Ibid.; Charles Francis Potter, "Ten Years After the Monkey Show I'm Going Back to Dayton," *Liberty*, 28 September 1935, p. 36.
113. Clarence Darrow, *The Story of My Life* (New York: Grosset, 1932), 251.

114. "Darrow Loud in His Protest," *Nashville Banner*, 8 July 1925, p. 1.

115. "Dayton Keyed Up," p. 1; Philip Kinsley, "Invoke Divine Guidance for Evolution Case," *Chattanooga Times*, 10 July 1925, p. 1.

Chapter Six. Preliminary Rounds

1. Sterling Tracy, "No Modernists Named on the Scopes Jury; All Believe in Bible," *Commercial Appeal* (Memphis), 11 July 1925, p. 1.

2. "Scopes Jury Chosen with Dramatic Speed," *New York Times*, 11 July 1925, p. 1.

3. "Dayton Disappointed," *Chattanooga Times*, 11 July 1925, p. 1.

4. Clarence Darrow, *The Story of My Life* (New York: Grosset, 1932), 256.

5. Arthur Garfield Hays, *Let Freedom Ring* (New York: Liveright, 1928), 34.

6. "Scopes Jury Chosen with Dramatic Speed," 1.

7. Ibid.; "Quash Indictment but Return New One," *Nashville Banner*, 10 July 1925, p. 1.

8. John T. Scopes, *Center of the Storm: Memoirs of John T. Scopes* (New York: Holt, 1967), 102–3.

9. W. J. Bryan to S. K. Hicks, 25 June 1925, in Hicks Papers.

10. Darrow, *My Life*, 256.

11. Scopes, *Center of the Storm*, 105; Hays, *Let Freedom Ring*, 76–77; Transcript, 3, 109.

12. Transcript, 41; Tracy, "No Modernists Named on the Scopes Jury," 1.

13. Transcript, 29–36.

14. Transcript, 13–14; Hays, *Let Freedom Ring*, 37; Watson Davis, "School of Science for Rhea County," *Chattanooga Times*, 11 July 1925, p. 1.

15. Transcript, 14.

16. H. L. Mencken, "The Monkey Trial: A Reporter's Account," in Jerry D. Tompkins, ed., *D-Days at Dayton: Reflections on the Scopes Trial* (Baton Rouge: Louisiana State University Press, 1965), 38–39 (reprint of 11 July 1925 article).

17. Tracy, "No Modernists Named on the Scopes Jury," 1.

18. "A Typical Southern Jury," *Pittsburgh American*, 17 July 1925, p. 4.

19. Transcript, 43.

20. Transcript, 7–9.

21. Davis, "School of Science for Rhea County," 1.

22. Transcript, 43.

23. Jack Lait, "Scopes Trial Keys Up Dayton," *Nashville Banner*, 12 July 1925, p. 1.

24. "Dayton's Police Suppress Skeptics," *New York Times*, 12 July 1925, p. 1.

25. Mencken, "Monkey Trial," 40 (reprint of 13 July 1925 article).

26. Ralph Perry, "'Loud Speaker' for the State," *Nashville Banner*, 13 July 1925, p. 1.

27. "Hostility Grows in Dayton Crowd; Champions Clash," *New York Times*, 12 July 1925, p. 1.

28. Ibid.
29. "Dayton's One Pro-Evolution Pastor Quits as Threat Bars Dr. Potter from Pulpit," *New York Times*, 13 July 1925, p. 1. (Article title refers to Howard G. Byrd, pastor of Dayton's northern Methodist church, who quit after his invitation to let Potter deliver the Sunday sermon was overruled by the congregation.)
30. "Dayton's One Pro-Evolution Pastor Quits," 1.
31. "Crowds Jam Court to See Champions," *New York Times*, 14 July 1925, p. 1.
32. Darrow, *My Life*, 259.
33. Transcript, 45.
34. Transcript, 50.
35. Transcript, 55.
36. Transcript, 56–57; Hays, *Let Freedom Ring*, 42–43.
37. "Clash of Attorneys," *Nashville Banner*, 13 July 1925, p. 3.
38. Transcript, 66.
39. Transcript, 66–68, 73.
40. "Lively Clashes in Move to Quash Indictment," *Chattanooga Times*, 14 July 1925, p. 1.
41. Darrow, *My Life*, 259–60.
42. "Darrow Scores Ignorance and Bigotry Seeking to Quash Scopes Indictment," *New York Times*, 14 July 1925, p. 1.
43. Transcript, 75.
44. Transcript, 79–84.
45. Transcript, 77–87.
46. "Darrow Scores Ignorance and Bigotry," 1.
47. Mencken, "Monkey Trial," 41 (reprint of 14 July 1925 article).
48. "Darrow Scores Ignorance and Bigotry," 1.
49. Philip Kinsley, "Liberty at Stake if Law Stands, Darrow Says," *Chicago Tribune*, 14 July 1925, p. 1.
50. Joseph Wood Krutch, "Darrow vs. Bryan," *Nation*, 29 July 1925, p. 136.
51. Hays, *Let Freedom Ring*, 46.
52. "Lively Clashes in Move to Quash Indictment," 1.
53. Darrow, *My Life*, 257.
54. "Darrow's Paradise," *Commercial Appeal* (Memphis), 15 July 1925, p. 1.
55. Mencken, "Monkey Trial," 41 (reprint of 14 July 1925 article).
56. Hays, *Let Freedom Ring*, 41.
57. Transcript, 89–90.
58. Transcript, 90–91; "Stormy Scenes in the Trial of Scopes as Darrow Moves to Bar All Prayers," *New York Times*, 15 July 1925, p. 1.
59. Transcript, 92.
60. Sterling Tracy, "Lawyers Out for Gore When Evolution Trial Starts to Get Rough," *Commercial Appeal* (Memphis), 15 July 1925, p. 1.
61. Transcript, 93–94; "Weird Adventures of 200 Reporters at Tennessee Evolution Trial," *Editor & Publisher*, 18 July 1925, p. 1.

62. Tracy, "Lawyers Out for Gore," 1.
63. Transcript, 97, 98, 102.
64. Scopes, *Center of the Storm*, 124.

Chapter Seven. The Trial of the Century

1. Ralph Perry, "Defense Calls Dr. Metcalf," *Nashville Banner*, 16 July 1925, p. 6.
2. John T. Scopes, *Center of the Storm: Memoirs of John T. Scopes* (New York: Holt, 1967), 138–39.
3. "Schoolboy Testimony State's Program Now," *Chattanooga Times*, 14 July 1925, p. 2.
4. Transcript, 112.
5. Transcript, 113–16.
6. Transcript, 113–15.
7. Sterling Tracy, "Scientific Evidence, Issue in Scopes Case, Is Pondered by Court," *Commercial Appeal* (Memphis), 16 July 1925, p. 1.
8. H. L. Mencken, "The Monkey Trial: A Reporter's Account," in Jerry D. Tompkins, ed., *D-Days at Dayton: Reflections on the Scopes Trial* (Baton Rouge: Louisiana State University Press, 1965), 44 (reprint of 14 July 1925 article).
9. Transcript, 121–22.
10. Transcript, 127.
11. Transcript, 129–31.
12. Transcript, 133.
13. Scopes, *Center of the Storm*, 136, 188.
14. Transcript, 138.
15. Mencken, "Monkey Trial," 45 (reprint of 7 July 1925 article).
16. Transcript, 139.
17. "Darrow Puts First Scientist on Stand to Instruct Scopes Judge on Evolution; State Completes Its Case in One Hour," *New York Times*, 16 July 1925, p. 1.
18. Sterling Tracy, "Malone Wins Cheers from Dayton People on Answering Bryan," *Commercial Appeal* (Memphis), 17 July 1925, p. 1.
19. Transcript, 153.
20. Transcript, 156–60.
21. Transcript, 163–64, 166–67, 169; "Prosecution Moves to Exclude Experts," *Nashville Banner*, 16 July 1925, p. 1.
22. Philip Kinsley, "They Call Us Bigots When We Refuse to Throw Away Our Bibles," *Commercial Appeal* (Memphis), 17 July 1925, p. 1 (*Chicago Tribune* wire story).
23. Transcript, 170.
24. Clarence Darrow, *The Story of My Life* (New York: Grosset, 1932), 263.
25. Transcript, 170–80.
26. Transcript, 180–82.

27. Scopes, *Center of the Storm*, 147–48.

28. Transcript, 184–88.

29. Sterling Tracy, "Malone Wins Cheers from Dayton People on Answering Bryan," *Commercial Appeal* (Memphis), 17 July 1925, p. 1; Arthur Garfield Hays, *Let Freedom Ring* (New York: Liveright, 1928), 65–66; John Washington Butler, "For Heaven's Sake!" *Commercial Appeal* (Memphis), 19 July 1925, p. 3; Scopes, *Center of the Storm*, 154–56.

30. Transcript, 190–99; Tracy, "Malone Wins Cheers," 1.

31. Ralph Perry, "Eventful Hour of Scopes Trial," *Nashville Banner*, 17 July 1925, p. 1.

32. "Bryan Defends Tennessee and Its Law; Calls Evolution Attack on Church; Spirited Debate on Expert Evidence," *New York Times*, 17 July 1925, p. 1.

33. Kinsley, "They Call Us Bigots," 1.

34. Transcript, 185.

35. Transcript, 201–3 (emphasis added); Scopes, *Center of the Storm*, 158–59; "Judge Shatters the Scopes Defense by Barring Testimony of Scientists; Sharp Clashes as Darrow Defies the Court," *New York Times*, 18 July 1925, p. 1.

36. Transcript, 204–9; Hays, *Let Freedom Ring*, 67–68.

37. Transcript, 207–9; Sterling Tracy, "Scientists Excluded, Darrow Spends a Day 'Strafing' the Judge," *Commercial Appeal* (Memphis), 18 July 1925, p. 1.

38. Mencken, "Monkey Trial," 50 (reprint of 18 July 1925 article).

39. "Weird Adventures of 200 Reporters at Tennessee Evolution Trial," *Editor & Publisher*, 18 July 1925, p. 1.

40. "Mencken Epithets Raise Dayton's Ire," *New York Times*, 17 July 1925, p. 3.

41. Charles Francis Potter, "Ten Years After the Monkey Show I'm Going Back to Dayton," *Liberty*, 28 September 1935, p. 37.

42. Bill Perry, "Scopes Defense Facing Defeat," *Nashville Banner*, 19 July 1925, p. 1.

43. "Defense Counsel at Work on Affidavits," *Nashville Banner*, 18 July 1925, p. 1; "Bryan's Statement," *Commercial Appeal* (Memphis), 19 July 1925, p. 1; John Herrick, "Bryan Stirs Up Animus Among Tennessee Folk," *Chicago Tribune*, 20 July 1925, p. 4.

44. "Offer of $10,000 to Start Bryan 'University' Opens Dayton Campaign for $1,000,000 Fund," *New York Times*, 17 July 1925, p. 1.

45. "Darrow's Statement," *Commercial Appeal* (Memphis), 19 July 1925, p. 1; "Bryan and Darrow Wage War of Words in Trial Interlude," *New York Times*, 19 July 1925, p. 1; "Bryan Now Regrets Barring of Experts," *New York Times*, 18 July 1925, p. 2.

46. Herrick, "Bryan Stirs Up Animus," 1.

47. "Defense Counsel Make Ready for Final Battle," *Nashville Tennessean*, 19 July 1925, p. 1.

48. Perry, "Scopes Defense Facing Defeat," 1.

49. Transcript, 211.

50. Transcript, 216–80.
51. Transcript, 225–27.
52. Scopes, *Center of the Storm*, 114.
53. "Bryan, Made Witness in Open Air Court, Shakes His Fist at Darrow Amid Cheers; Apology End Contempt Proceedings," *New York Times*, 21 July 1925, p. 1.
54. "Defense Counsel Make Ready for Final Battle," 1.
55. Scopes, *Center of the Storm*, 166.
56. Transcript, 288.
57. "Big Crowd Watches Trial Under Trees," *New York Times*, 21 July 1925, p. 1.
58. Ralph Perry, "Added Thrill Given Dayton," *Nashville Banner*, 21 July 1925, p. 2.
59. Darrow, *My Life*, 267.
60. Transcript, 285.
61. Transcript, 302.
62. Hays, *Let Freedom Ring*, 77.
63. Scopes, *Center of the Storm*, 178.
64. Transcript, 302.
65. "Bryan, Made Witness in Open Air Court," 1.
66. Transcript, 299.
67. Transcript, 304.
68. Ralph Perry, "Added Thrill Given Dayton," *Nashville Banner*, 21 July 1925, p. 2.
69. Sterling Tracy, "Darrow Quizzes Bryan; Agnosticism in Clash with Fundamentalism," *Commercial Appeal* (Memphis), 21 July 1925, p. 1.
70. Clarence Darrow to H. L. Mencken, 15 August 1925, in H. L. Mencken Collection, New York Public Library, NY.
71. Transcript, 305.
72. Transcript, 306–8.
73. Transcript, 311–12.
74. Corinne Rich, "Jurors Know Least About Scopes Trial," *Commercial Appeal* (Memphis), 22 July 1925, p. 1.
75. "Scopes Fined $100," *Chattanooga Times*, 22 July 1925, p. 1.
76. Sterling Tracy, "Scopes Is Convicted; Draws $100 Fine for Teaching Evolution," *Commercial Appeal* (Memphis), 22 July 1925, p. 1.
77. Transcript, 316–17.

Chapter Eight. The End of an Era

1. Lawrence W. Levine, *Defender of the Faith: William Jennings Bryan, The Last Decade, 1915–1925* (New York: Oxford University Press, 1965), 355.
2. "Commoner Propounds 9 Specific Questions to Chicago Attorney," *Knoxville Journal*, 22 July 1925, p. 8.

3. "Dayton Hears Parting Shots," *Nashville Banner*, 22 July 1925, p. 4.

4. "Bryan Doesn't Claim 'To Know Everything'; He Replies to Foes," *Commercial Appeal* (Memphis), 23 July 1925, p. 1.

5. Transcript, 338. (Bryan's unused closing argument was printed as a supplement in the unofficial published version of the trial transcript.)

6. George F. Milton, "A Dayton Postscript," *Outlook* 140 (1925), 552. For a similar comment, see Milton's editorial, "Disgraceful Performance," *Chattanooga News*, 21 July 1925, p. 8.

7. Ray Ginger, *Six Days or Forever? Tennessee v. John Thomas Scopes* (London: Oxford University Press, 1958), 192; "Bryan Satisfied with His Recent Crusades," *Commercial Appeal* (Memphis), 23 July 1925, p. 3.

8. Grace Dexter Bryan to Judge Sue Hicks, 12 April 1940, in Hicks Papers.

9. William Jennings Bryan and Mary Baird Bryan, *The Memoirs of William Jennings Bryan* (Philadelphia: United, 1925), 485–86.

10. Transcript, 339.

11. Irving Stone, *Clarence Darrow for the Defense* (New York: Doubleday, 1941), 464; Joseph Wood Krutch, "The Great Monkey Trial," *Commentary* (May 1967), 84; Robert D. Linder, "Fifty Years After Scopes: Lessons to Learn, a Heritage to Reclaim," *Christianity Today*, 18 July 1975, p. 9.

12. Arthur Garfield Hays, *Let Freedom Ring* (New York: Liveright, 1928), 79–80.

13. "Dayton Snap Shots," *Nashville Banner*, 12 July 1925, p. 8.

14. "Dayton Back to Earth," *Commercial Appeal* (Memphis), 23 July 1925, p. 2; "Dayton's Subsidence," *Nashville Banner*, 22 July 1925, p. 8; John T. Scopes, *Center of the Storm: Memoirs of John T. Scopes* (New York: Holt, 1967), 191–95.

15. Scopes, *Center of the Storm*, 194, 206–7.

16. Russell D. Owen, "The Significance of the Scopes Trial," *Current History* 22 (1925), 875.

17. Herbert E. Hicks to Ira Evans Hicks, 22 July 1925, in Hicks Papers; "Malone Talks at Follies," *New York Times*, 24 July 1925, p. 13; Arthur Garfield Hays, "The Strategy of the Scopes Defense," *The Nation*, 5 August 1925, p. 158.

18. H. L. Mencken, "The Monkey Trial: A Reporter's Account," in Jerry D. Tompkins, ed., *D-Days at Dayton: Reflections on the Scopes Trial* (Baton Rouge: Louisiana State University Press, 1965), 51 (reprint of 18 July 1925 article); "Says Evolution Laws Will Become General," *Commercial Appeal* (Memphis), 23 July 1925, p. 4.

19. Ralph Perry, "Both Won in Scopes Hearing," *Nashville Banner*, 22 July 1925, p. 1.

20. T. W. Callaway, "Think Darrow Met His Match," *Chattanooga Times*, 22 July 1925, p. 2; W. S. Keese, "Declares Bryan Shorn of Strength," *Chattanooga Times*, 22 July 1925, p. 2.

21. "Real Religion and Real Science," *Commercial Appeal* (Memphis), 26 July 1925, sec. 1, p. 4.

22. Frank R. Kent, "On the Dayton Firing Line," *The New Republic* 43 (1925), 259.

23. "Ended at Last," *New York Times*, 22 July 1925, p. 18; "As Expected, Bryan Wins," *Chicago Tribune*, 22 July 1925, p. 8.

24. "Dayton's 'Amazing' Trial," *Literary Digest*, 25 July 1925, p. 7.

25. "2,000,000 Words Wired to the Press," *New York Times*, 22 July 1925, p. 22; "The End in Sight at Dayton," *New York Times*, 18 July 1925, p. 12; Transcript, 316.

26. Howard W. Odum, "Duel to the Death," *Social Forces* 4 (1925), 190.

27. "His Death Dramatic," *New York World*, 27 July 1925, p. 16.

28. Scopes, *Center of the Storm*, 203; G. W. Rappleyea to Forrest Bailey, 7 August 1925, in ACLU Archives, vol. 274; Austin Peay, "The Passing of William Jennings Bryan," in Austin Peay, *A Collection of State Papers and Political Addresses* (Kingsport, TN: Southern, 1929), 450.

29. "Comment of Press of Nation on Bryan's Death," *New York Times*, 27 July 1925, p. 2.

30. Charles O. Oaks, "Death of William Jennings Bryan," in Norm Cohen, "Scopes and Evolution in Hillbilly Songs," *JEMF Quarterly* 6 (1970), 176; W. B. Riley, "Bryan: The Great Commoner and Christian," *Christian Fundamentals in School and Church* 7 (October 1925), 9, 11.

31. "Evolution Issue in Congress, Forecast," *Commercial Appeal* (Memphis), 30 July 1925, p. 1.

32. "Mississippi May Ban Theory of Evolution," *Commercial Appeal* (Memphis), 31 July 1925, p. 1; "Tennessee Man Attacks Evolution," *Clarion-Ledger* (Jackson), 9 February 1926, p. 3.

33. Frank R. Kent, "On the Dayton Firing Line," *The New Republic* 43 (1925), 260; "Dr. John Roach Straton Challenges Darrow," *Johnstown Democrat*, 20 August 1925, p. 16; Riley, "Bryan: The Great Commoner and Christian," 11. About the same time, J. Frank Norris compared Bryan standing against Darrow to "Moses challenging Pharaoh" and "Martin Luther hurling his thesis [*sic*] at Pope Leo X. It is the greatest battle of the centuries." See James J. Thompson, Jr., *Trial as by Fire: Southern Baptists and the Religious Controversies of the 1920s* (Macon, Ga.: Mercer University Press, 1982), 132.

34. Cohen, "Scopes and Evolution in Hillbilly Songs," 176–81; "Demand for Special Record," *Talking Machine World* (15 September 1925), 83; see also Mel R. Wilhoit, "Music of the Scopes Monkey Trial," typescript, Bryan College Music Department, Dayton, Tennessee, 1995. The country music classic, "A Boy Named Sue," a distant cousin of these Scopes songs, was inspired by the Scopes prosecutor Sue Hicks. Juanita Glenn, "Judge Still Recalls 'Monkey Trial'—50 Years Later," *Knoxville Journal*, 11 July 1975, 17.

35. H. L. Mencken, "Editorial," *American Mercury* 6 (1925), 159.

36. Ronald L. Numbers, "The Scopes Trial: History and Legend," *Southern Culture* (forthcoming); "Dayton and After," *Nation* 121 (1925), 155–56; Mencken, "Editorial," 160; Maynard Shipley, *The War on Modern Science* (New York: Knopf, 1927), 3–4.

37. "Darrow's Blunder," *New York World*, 23 July 1925, p. 18; "Darrow Betrayed Himself," *Times-Picayune* (New Orleans), 23 July 1925, p. 8.

38. "The Scopes Case Counsel," *Religious Weekly Review*, 27 June 1925, in ACLU Archives, vol. 276; Brower Eddy to John R. Neal, 10 September 1925, in ACLU Archives, vol. 274; Edwin Mims, "Modern Education and Religion," 6, in Mims Papers; Raymond B. Fosdick to Roger N. Baldwin, 19 October 1925, in ACLU Archives, vol. 274; Roger N. Baldwin to Raymond B. Fosdick, 21 October 1925, in ACLU Archives, vol. 274.

39. ACLU Executive Committee, "Minutes," 3 August 1925, in ACLU Archives, vol. 279; Forrest Bailey to Clarence Darrow, 2 September 1925, in ACLU Archives, vol. 275 (quoting from earlier letter to Neal); Rappleyea to Bailey, 7 August 1925.

40. Forrest Bailey to Charles H. Strong, 12 August 1925, in ACLU Archives, vol. 274 (similar letters in same volume); "The Conduct of the Scopes Trial," *The New Republic* 43 (1925), 332.

41. Bailey to Darrow, 2 September 1925; Clarence Darrow to Forrest Bailey, 4 September 1925, in ACLU Archives, vol. 274.

42. Forrest Bailey to Walter Nelles, 4 September 1925, in ACLU Archives, vol. 274; "Mr Hughes and the Tennessee Law," *New York World*, 3 September 1925, p. 8; Arthur Garfield Hays to Walter Nelles, 9 September 1925, in ACLU Archives, vol. 274; Walter Nelles to Arthur Garfield Hays, 10 September 1925, in ACLU Archives, vol. 274.

43. Transcript, 288; Mencken, "Monkey Trial," 51 (reprint of 18 July 1925 article); Joseph Wood Krutch, "Darrow vs. Bryan," *Nation,* 29 July 1925, p. 136; Maynard M. Metcalf et al. to Michael I. Pupin, 17 August 1925, in Darrow Papers.

44. Forrest Bailey to John T. Scopes c/o Clarence Darrow, 29 September 1925, in ACLU Archives, vol. 274.

45. Clarence Darrow to Forrest Bailey, 10 February 1926, in ACLU Archives, vol. 299; Forrest Bailey to Franklin Reynolds, 23 December 1925, in ACLU Archives, vol. 274. For an example of local attorneys handling matters, see Franklin Reynolds to Forrest Bailey, 10 December 1925, in ACLU Archives, vol. 274.

46. See, e.g., Forrest Bailey to Walter Nelles, 4 September 1925, in ACLU Archives, vol. 274; Forrest Bailey to Frank H. O'Brien, 3 December 1925, in ACLU Archives, vol. 274.

47. K. T. McConnico to Charles L. Cornelius, 16 September 1926, in Peay Papers, GP 40–24.

48. Austin Peay to Samuel Untermyer, 19 September 1925, in Peay Papers, GP 40–24.

49. "Condensed Minutes of Annual Meeting," *Journal of the Tennessee Academy of Sciences* 1 (1925), 9; Wilson L. Newman to Austin Peay, 5 December 1925, in Peay Papers, GP 40–13.

50. For the correspondence to Peay on this matter, see Peay Papers, GP 40–13,

which also includes a *Nashville Banner* article summarizing the letters. For the ACLU offer to support a challenge to the Mississippi law, see American Civil Liberties Union, "Press Service," 20 March 1926, in ACLU Archives, vol. 299.

51. George F. Milton to Austin Peay, 8 August 1925, in Peay Papers, GP 40–24; Franklin Reynolds to Forrest Bailey, 18 March 1926, in ACLU Archives, vol. 299; John T. Scopes to Roger N. Baldwin, 8 August 1926, in ACLU Archives, vol. 299.

52. Forrest Bailey to Arthur Garfield Hays, 5 January 1926, in ACLU Archives, vol. 299; Forrest Bailey to Robert S. Keebler, 5 January 1926, in ACLU Archives, vol. 299; Arthur Garfield Hays to Forrest Bailey, 6 January 1926, in ACLU Archives, vol. 299; Robert S. Keebler to Forrest Bailey, 9 February 1926, in ACLU Archives, vol. 299; Arthur Garfield Hays to Walter Nelles, 9 September 1925, in ACLU Archives, vol. 274. Bailey miscounted the number of Tennessee and non-Tennessee lawyers on the defense brief. There were five "foreigners" (Darrow, Malone, Hays, Rosensohl, and the ACLU attorney Walter H. Pollak), and four "natives" (Neal, Keebler, Spurlock, and a local attorney named Frank McElwee, who had advised the defense throughout the trial and appeal).

53. John Randolph Neal to Forrest Bailey, 15 February 1926, in ACLU Archives, vol. 299.

54. "Reply Brief and Argument for the State of Tennessee," State v. Scopes, 154 Tenn. 105 (1926), pp. 14, 78–80, 380 (emphasis in original).

55. "Brief and Argument of the Tennessee Academy of Sciences as Amicus Curiae," Scopes v. State, 154 Tenn. 105 (1926), pp. 16, 90, 154.

56. "World Awaits Scopes Hearing Here Monday," *Nashville Banner*, 30 May 1926, p. 1; "State Defends Anti-Evolution Law," *Knoxville Journal*, 1 June 1926, p. 1.

57. "Supreme Court Hears Scopes Case," *Nashville Banner*, 31 May 1926, p. 1.

58. Ibid.; "Anti-Evolution Law Called 'Capricious'," *Commercial Appeal* (Memphis), 1 June 1926, p. 1.

59. "Supreme Court Hears Scopes Case," 1.

60. William Hutchinson, "Darrow Makes Fervid Plea," *Nashville Banner*, 1 June 1926, p. 1.

61. "Scopes Case Rests in Hands of State's Highest Tribunal," *Knoxville Journal*, 2 June 1926, p. 1; "Darrow and McConnico Speak in Scopes Case," *Nashville Banner*, 1 June 1926, p. 1.

62. Ibid.; "Darrow Declares Science as Real as Religion," *Chattanooga Times*, 2 June 1926, p. 1.

63. "Argument of Clarence Darrow," Scopes v. State, 154 Tenn. 105 (1926), pp. 17, 26–28, in Darrow Papers; Hutchinson, "Darrow Makes Fervid Plea," 1; Hays, *Let Freedom Ring*, 80.

64. Hays, *Let Freedom Ring*, 80; "Religious Issue Flares in Scopes Case Pleas," *Chattanooga Times*, 1 June 1926, p. 1; "Scopes Case," 1 (Associated Press wire story).

65. Forrest Bailey to Clarence Darrow, 3 June 1926, in ACLU Archives, vol. 299; Clarence Darrow to Forrest Bailey, 9 June 1926, in ACLU Archives, vol. 299; Roger N. Baldwin to John T. Scopes, 10 August 1926, in ACLU Archives, vol. 299; Wolcott H. Pitkin to Felix Frankfurter, in ACLU Archives, vol. 299.

66. Scopes, *Center of the Storm*, 237.

67. Scopes v. State, 154 Tenn. 105, 289 S.W. at 363, 364, 367, 370 (1927).

68. Ibid., 289 S.W. at 367.

69. "Scopes Goes Free, but Law Is Upheld," *New York Times*, 16 January 1927, p. 1; "Will Ask Court to Rehear Case," *Nashville Banner*, 17 January 1927, p. 1.

70. "Finis Is Written in Scopes Case," *Nashville Banner*, 16 January 1927, p. 1.

71. Lida B. Robertson to Governor Peay, 11 August 1925, in Peay Papers, GP 40–13.

72. Shipley, *War on Science*, 111.

73. See, e.g., Virginia Gray, "Anti-Evolution Sentiment and Behavior: The Case of Arkansas," *Journal of American History* 62 (1970), 357–65.

74. "Malone Talks," 13; "The Inquisition in Tennessee," *The Forum* 74 (1925), 159; Edwin Mims, "Mr. Mencken and Mr. Sherman: Smartness and Light," in Mims Papers, box 19; Edward J. Larson, *Trial and Error: The American Controversy Over Creation and Evolution* (New York: Oxford University Press, 1989), 83–84.

75. Hays, "Strategy of the Scopes Defense," 157; Clarence Darrow, *The Story of My Life* (New York: Grosset, 1932), 267. See also Hays, *Let Freedom Ring*, 79; Arthur Garfield Hays, *City Lawyer* (New York: Simon & Schuster, 1942), 215.

76. W. B. Riley, "The World's Christian Fundamentals Association and the Scopes Trial," *Christian Fundamentals in School and Church* 7 (October 1925), 39–40.

77. Charles A. Beard and Mary R. Beard, *The Rise of American Civilization*, vol. 2 (New York: Macmillan, 1928), 752–53.

78. Preston William Slosson, *The Great Crusade and After* (New York: Macmillan, 1931), 432–33. For a similar account, see William W. Sweet, *The Story of Religion in America* (New York: Harper, 1930), 513.

79. Paxton Hibbon, *Peerless Leader: William Jennings Bryan* (New York: Farrar, 1929), 402 (quote); Morris R. Werner, *Bryan* (New York: Harcourt, 1929), 339–55.

80. For an extended analysis of this issue, see Paul M. Waggoner, "The Historiography of the Scopes Trial: A Critical Re-evaluation," *Trinity Journal* (new series), 5 (1985), 161.

Chapter Nine. Retelling the Tale

1. Frederick Lewis Allen, *Only Yesterday: An Informal History of the Nineteen-Twenties* (reprint, New York: Harper, 1964), 163–71.

2. Ibid., 163–64, 170; Clarence Darrow, *The Story of My Life* (New York: Grosset, 1932), 267.

3. Allen, *Only Yesterday*, vii–viii.

4. Roderick Nash, *The Nervous Generation: American Thought, 1917–1930* (Chicago: Rand McNally, 1970), 5–8. See also Darwin Payne, *The Making of Only Yesterday: Frederick Lewis Allen* (New York: Harper, 1975), 98–103.

5. Ernst Mayr, personal communication, 1 December 1995.

6. Allen, *Only Yesterday*, 168–70; Paul M. Waggoner, "The Historiography of the Scopes Trial: A Critical Re-evaluation," *Trinity Journal* (new series), 5 (1985), 161.

7. Gaius Glen Atkins, *Religion in Our Times* (New York: Round Table, 1932), 250–52; Mark Sullivan, *Our Times: The United States, 1900–1925* (New York: Scribner's, 1935), 644.

8. William W. Sweet, *The Story of Religion in America* (New York: Harper, 1930), 513; William W. Sweet, *The Story of Religion in America*, rev. ed. (New York: Harper, 1939); Irving Stone, *Clarence Darrow for the Defense* (Garden City: Doubleday, 1941), 437.

9. W. J. Bryan to Dr. Howard A. Kelly, 17 June 1925, in Bryan Papers; John Thomas Scopes to Editor, *Forum* (June 1925), xxvi.

10. William Vance Trollinger, Jr., "Introduction," in William Vance Trollinger, Jr., ed., *The Antievolution Pamphlets of William Bell Riley* (New York: Garland, 1995), xvii–xviii.

11. Howard W. Odum, *An American Epoch: Southern Portraiture in the National Picture* (New York: Holt, 1930), 167–68. For a similar comment later in the decade, see Howard W. Odum, *Southern Regions of the United States* (Chapel Hill: University of North Carolina Press, 1936), 501, 527.

12. C. H. Thurber to W. J. Bryan, 21 November 1923, in Bryan Papers; W. J. Bryan to C. H. Thurber, 22 December 1923, in Bryan Papers.

13. Compare George William Hunter, *A Civic Biology* (New York: American, 1914), 193–96, 235, 404–6, 423, with George William Hunter, *A New Civic Biology* (New York: American, 1926), 250, 383, 411–12, 436. For a broad analysis of multiple texts, see Judith V. Grabner and Peter D. Miller, "Effect of the Scopes Trial," *Science* 185 (1974), 832–37; Gerald Skoog, "The Topic of Evolution in Secondary School Biology Textbooks: 1900–1977," *Science Education* 63 (1979), 620–36; Edward J. Larson, *Trial and Error: The American Controversy Over Creation and Evolution* (New York: Oxford University Press, 1989), 84–88.

14. Ronald L. Numbers, *The Creationists: The Evolution of Scientific Creationism* (New York: Knopf, 1992), 100; Edward B. Davis, "Introduction," in Edward B. Davis, ed., *The Antievolution Pamphlets of Harry Rimmer* (New York: Garland, 1995), xvi–xix.

15. George M. Marsden, *Fundamentalism and American Culture: The Shaping of Twentieth-Century Evangelicalism, 1870–1925* (New York: Oxford University Press, 1980), 184–85.

16. Joel A. Carpenter, "Fundamentalist Institutions and the Rise of Evangelical Protestantism, 1929–1942," *Church History* 49 (1980), 62–75 (Carpenter quote from Marsden, *Fundamentalism and American Culture*, 194).

17. Ronald L. Numbers, "The Creationists," in Martin E. Marty, ed., *Fundamentalism and Evangelicalism* (Munich: Saur, 1993), 261.

18. John 18:36 (AV); 1 Cor. 1:20 (AV).

19. "I'll Fly Away," in *Wonderful Melody* (Hartford, Conn.: Hartford Music, 1932).

20. Richard Hofstadter, *The American Political Tradition: And the Men Who Made It* (New York: Knopf, 1948), 199–202. For later, more balanced presentations, see Lawrence W. Levine, *Defender of the Faith: William Jennings Bryan, The Last Decade, 1915–1925* (New York: Oxford University Press, 1965); Paolo E. Coletta, *William Jennings Bryan.* Vol. 3. *Political Puritan, 1915–1925* (Lincoln: University of Nebraska Press, 1969).

21. Norman F. Furniss, *The Fundamentalist Controversy, 1918–1931* (New Haven, Conn.: Yale University Press, 1954), 3.

22. William E. Leuchtenburg, *The Perils of Prosperity, 1914–1932* (Chicago: University of Chicago Press, 1958), 217–23.

23. Ray Ginger, *Six Days or Forever? Tennessee v. John Thomas Scopes* (London: Oxford University Press, 1958), 190–217, 238.

24. Richard Hofstadter, *The Age of Reform: From Bryan to F.D.R.* (New York: Knopf, 1955), 286.

25. Richard Hofstadter, William Miller, and Daniel Aaron, *The United States: The History of a Republic* (Englewood Cliffs, N.J.: Prentice-Hall, 1957), 636; Irwin Unger, *These United States: The Questions of Our Past*, vol. 2, 6th ed. (Englewood Cliffs, N.J.: Prentice-Hall, 1995), 712; Samuel Eliot Morison, Henry Steele Commanger, and William E. Leuchtenburg, *A Concise History of the American Republic*, rev. ed. (New York: Oxford University Press, 1977), 588; William Miller, *A New History of the United States* (New York: Braziller, 1958), 356. A half-dozen other collegiate textbooks published between 1960 and 1995 contain similar accounts of the trial.

26. See Michael Lienesch, *Redeeming America: Piety and Politics in the New Christian Right* (Chapel Hill: University of North Carolina Press, 1993), 154.

27. Compare Harry Rimmer, "The Theories of Evolution and the Facts of Human Antiquity" (1929), in Davis, ed., *Antievolution Pamphlets*, 84–85, with later pamphlets in this collection; compare with Arthur I. Brown, "Science Speaks to Osborn," in Ronald L. Numbers, ed., *The Antievolution Works of Arthur I. Brown* (New York: Garland, 1995), 134–75, with other works in this collection.

28. George McCready Price to William Jennings Bryan, 1 July 1925, in Bryan Papers; "Says Millions Here Oppose Darwinism," *New York Times*, 8 September 1925, p. 9; George McCready Price, "What Christians Believe About Creation," *Bulletin of Deluge Geology* 2 (1942), 76.

29. Henry M. Morris, *History of Modern Creationism* (San Diego: Master, 1984), 73.

30. Jerry Falwell, *The Fundamentalist Phenomenon: The Resurgence of Conservative Christianity* (Garden City: Doubleday, 1981), 86.

31. "House Decides State to Keep Evolution Act," *Chattanooga Times*, 20 February 1935, p. 2.

32. Judson A. Rudd to Members of the Legislature, 15 March 1951, in Scopes trial file, Bryan College Archives, Dayton, Tennessee.

33. Ibid. For a discussion of this period of anti-Communist activity by fundamentalists, see James Davison Hunter, *Evangelicalism: The Coming Generation* (Chicago: University of Chicago Press, 1987), 121–24.

34. Ferenc M. Szasz, "William Jennings Bryan, Evolution and the Fundamentalist–Modernist Controversy," in Marty, ed., *Fundamentalism and Evangelicalism*, 109. For an example of a later defense of Bryan, see Robert D. Linder, "Fifty Years After Scopes: Lessons to Learn, a Heritage to Reclaim," *Christianity Today*, 18 July 1975, pp. 7–10.

35. Hunter, *Evangelicalism*, 120.

36. Ginger, *Six Days or Forever?* 238.

37. Richard Hofstadter, *Anti-Intellectualism in American Life* (New York: Knopf, 1963), 3, 125, 130–31.

38. Jerome Lawrence and Robert E. Lee, "Inherit the Wind: The Genesis & Exodus of the Play," *Theater Arts* (August 1957), 33; Elizabeth J. Haybe, "A Comparison Study of *Inherit the Wind* and the Scopes 'Monkey Trial'," Master's thesis, University of Tennessee, 1964, p. 66; Tony Randall, personal communication, April 1996.

39. Lawrence and Lee, "Genesis & Exodus," 33.

40. Jerome Lawrence and Robert E. Lee, *Inherit the Wind* (New York: Bantam, 1960), vii, 4; John T. Scopes, *The Center of the Storm: Memoirs of John T. Scopes* (New York: Holt, 1967), 270.

41. Lawrence and Lee, *Inherit the Wind*, 3, 7, 64.

42. Joseph Wood Krutch, "The Monkey Trial," *Commentary* (May 1967), 84.

43. Lawrence and Lee, *Inherit the Wind*, 7, 30, 63.

44. Ronald L. Numbers, "Inherit the Wind," *Isis* 84 (1993), 764.

45. Lawrence and Lee, *Inherit the Wind*, 85, 91, 103; Gerald Gunther, personal communication, 17 November 1995.

46. Lawrence and Lee, *Inherit the Wind*, 32, 42.

47. Ibid., 112–15.

48. Andrew Sarris, "Movie Guide," *Village Voice*, 10 November 1960, p. 11.

49. "Mixed Bag," *The New Yorker*, 30 April 1955, p. 67; "The New Pictures," *Time*, 17 October 1960, p. 95; Robert Hayes, "Our American Cousin," *Commonweal* 62 (1955), 278; Walter Kerr, "Inherit the Wind," *New York Herald Tribune*, 22 April 1955, p. 10; Stanley Kauffmann, "O Come All Ye Faithful," *The New Republic*, 31 October 1960, pp. 29–30; Sarris, "Movie Guide," 11. The movie version did not win any academy awards, although Tracy was nominated as Best Actor for his portrayal of Drummond-Darrow. Interestingly, he

lost to Burt Lancaster, who won it for his performance in the title role of *Elmer Gantry*, a part loosely based on the antievolutionist pastor John Roach Straton.

50. Scopes, *Center of the Storm*, 210; Juanita Glenn, "Judge Still Recalls 'Monkey Trial'—50 Years Later," *Knoxville Journal*, 11 July 1975, p. 17. As a fiercely loyal Democratic politician, Bryan would have had one consolation: the release of the movie coincided with the 1960 presidential election, and its thinly veiled attack on McCarthyism may have contributed to the narrow defeat of the Republican party's red-baiting nominee, Richard M. Nixon.

51. Krutch, "Monkey Trial," 83.

52. Ibid.; National Center for History in Schools, *National Standards for United States History: Exploring the American Experience* (Los Angeles: National Center, [1994]), 180.

53. Carl Sagan, personal communication, 21 November 1995; Howard J. Van Till, personal communication, 27 December 1995.

54. Morris, *History of Modern Creationism*, 76–77. For examples of these attempts, see David N. Menton, "Inherit the Wind: A Hollywood History of the Scopes Trial," *Contrast* (January 1985), p. 1; Euphemia Van Rensselaer Wyatt, "Theater," *Catholic World* 181 (1955), 226; Phillip E. Johnson, *Darwin on Trial* (Washington, D.C.: Regnery, 1991), 4–6.

55. Martin E. Marty, *Righteous Empire: The Protestant Experience in America* (New York: Dial, 1970), 220; Martin E. Marty, personal communication, 29 November 1995.

56. Stephen Jay Gould, *Hen's Teeth and Horse's Toes* (New York: Norton, 1983), 270, 273.

57. Allen, *Only Yesterday*, 171; Hofstadter, *Anti-Intellectualism*, 129; Lawrence and Lee, *Inherit the Wind*, 109; Randall, personal communication, April 1996; Kauffman, "O Come All Ye Faithful," 29.

Chapter Ten. Distant Echoes

1. "The Conduct of the Scopes Trial," *The New Republic* 43 (1925), 332.

2. Gitlow v. New York, 208 U.S. 652 (1925)(free speech); Everson v. Board of Education, 330 U.S. 1 (1947)(establishment clause); McCullum v. Board of Education, 333 U.S. 203 (1948)(religious instruction); Engel v. Vitale, 370 U.S. 421 (1962)(school prayer); Abington School Dist. v. Shempp, 374 U.S. 203 (1963)(Bible reading). Following these decisions, the ACLU annual report stated, "We are confident that when more sectarian religious practices are brought to the Court's attention, they likewise will be declared unconstitutional." American Civil Liberties Union, *Freedom Through Dissent: 42nd Annual Report* (New York: ACLU, [1963]), 22.

3. For discussions of this development, see Arnold B. Grobman, *The Changing Classroom: The Role of the Biological Sciences Curriculum Study* (Garden City: Doubleday, 1969), 94–95, 204; Gerald Skoog, "The Topic of Evolu-

tion in Secondary School Biology Textbooks: 1900–1977," *Science Education* 63 (1979), 632–33.

4. Bud Lumke, "Science Teacher Takes Stand in Evolution Hearing," *Arkansas Democrat*, 1 April 1966, p. 1; "Proceedings," in Appendix, Epperson v. Arkansas, 393 U.S. 97 (1968), 40–60; "Teacher Fired on Evolution," *Knoxville Journal*, 15 April 1967, p. 1.

5. "The Press-Scimitar Blitzes the Tennessee Anti-Evolution Law," *Scripps-Howard News* (August 1967), 9; "Monkey Law Bill May Be Decided," *Nashville Tennessean*, 10 May 1967, p. 8.

6. Lorry Daughtrey, "House Act Fails to Stir Scopes," *Nashville Tennessean*, 13 April 1967, p. 1.

7. "Press-Scimitar Blitzes Tennessee Anti-Evolution Law," 9.

8. "House Votes Down 'Monkey Law'," *Nashville Banner*, 12 April 1967, p. 1; "I May Be Leaving," *Nashville Tennessean*, 15 April 1967, p. 4 (editorial cartoon); Daughtrey, "House Act Fails to Stir Scopes," 1.

9. Bill Kovach, "'Monkey Law' Left Out on a Limb," *Nashville Tennessean*, 21 April 1967, p. 1; "Seems I'm Still Here," *Nashville Tennessean*, 22 April 1967, p. 4.

10. "'Monkey Law' Vote Stalled," *Nashville Tennessean*, 12 May 1967, p. 12; "Rehired Teacher to Test 'Monkey Law' Anyway," *Nashville Tennessean*, 13 May 1927, p. 1; "Overthrow of Monkey Law Asked," *Knoxville Journal*, 16 May 1967, p. 3; "Anti-Evolution Law Brings Shame on State," *Nashville Tennessean*, 15 May 1976, p. 8; William Bennett, "State's 'Monkey Law' Repealed by Senators," *Commercial Appeal* (Memphis), 17 May 1967, p. 1; "Scott to End Suit; Scopes Welcomes Action in Assembly," *Nashville Banner*, 17 May 1967, p. 2; Walter Smith, "Monkey Law Dead, but Dayton Residents Recall Famed Trial," *Nashville Banner*, 19 May 1967, p. 14 (national wire service article).

11. State v. Epperson, 242 Ark. 922, 416 S.W.2d 322, 322 (1967).

12. "In Court's Failure, the Barrier Remains," *Arkansas Gazette*, 8 June 1967, p. 6A.

13. Peter L. Zimroth, "Epperson and Blanchard v. Ark.," 20 December 1967, in Fortas Papers; Peter L. Zimroth, "Supp. Memo," 16 February 1968, in Fortas Papers.

14. Arthur Goldschmidt to Abe Fortas, 22 November 1968, in Fortas Papers.

15. Louis R. Cohen, "Epperson v. Arkansas," 14 December 1967, p. 3, in John Marshall Harlan Papers, Princeton University Library; "Brief for Appellants," Epperson v. Arkansas, 393 U.S. 97, p. 8; "Brief for Appellee," Epperson v. Arkansas, 393 U.S. 97, pp. 1, 28–31; "Brief of American Civil Liberties Union and American Jewish Congress," Epperson v. Arkansas, 393 U.S. 97, p. 2; for example (Transcript of Oral Arguments), Epperson v. Arkansas, 393 U.S. 97, p. 14.

16. "No. 7, Epperson v. Arkansas," 18 October 1968, in Fortas Papers. All but acknowledging the statute's lack of religious effect, Fortas wrote in the initial handwritten draft of his opinion, "For our purposes, it is unimportant that the

theory of evolution continues to live and to command substantial adherence, probably even in Arkansas' publicly supported institutions of learning." In Epperson case file, 26 October 1968, in Fortas Papers.

17. Epperson v. Arkansas, 393 U.S. at 102–9.

18. Lemon v. Kurtzman, 403 U.S. 602, 612 (1971); Gerald Gunther, personal communication, 17 November 1995; Charles Alan Wright, personal communication, 21 November 1995.

19. Epperson v. Arkansas, 393 U.S. at 239 (Black, J., concurring).

20. Epperson v. Arkansas, 393 U.S. at 109. For the added language, compare the published opinion with Fortas's handwritten draft dated 26 October 1968, in Fortas Papers.

21. "Court Rules in 'Scopes Case'," *U.S. News and World Report*, 22 November 1968, p. 16; "Making Darwin Legal," *Time*, 22 November 1968, p. 41; "Evolution Revolution in Arkansas," *Life*, 22 November 1968, p. 89; Fred P. Graham, "Court Ends Darwinism Ban," *New York Times*, 12 November 1968, p. 1.

22. Jerome Lawrence and Robert E. Lee, *Inherit the Wind* (New York: Bantam, 1960), 89; Epperson v. Arkansas, 393 U.S. at 109. For a typical Bryan plea for neutrality, see William Jennings Bryan, "God and Evolution," *New York Times*, 26 February 1922, sec. 7, p. 1.

23. Transcript, 185–87.

24. Epperson v. Arkansas, 393 U.S. at 109.

25. Wendell R. Bird, "Freedom from Establishment and Unneutrality in Public School Instruction and Religious School Regulation," *Harvard Journal of Law and Public Policy* 2 (1979), 179.

26. This version of the quote—and there are several in various creationist writings, but none with any authoritative reference to Darrow himself—appeared as the introduction to Wendell R. Bird, "Creation-Science and Evolution-Science in Public Schools: A Constitutional Defense in Public Schools," *Northern Kentucky Law Review* 9 (1982), 162.

27. George Gallup, "Public Evenly Divided Between Evolutionists, Creationists," *Los Angeles Times Syndicate*, 1982, p. 1 (press release); "76% for Parallel Teaching of Creation Theories," *San Diego Union*, 18 November 1981, p. A15 (reporting national poll where over 80 percent of respondents favored either parallel teaching or exclusive teaching of creationism).

28. Tenn. Code Ann. sec. 49–2008; Ark. Stat. Ann. sec. 80–1663, *et. sec.* (1981 Supp.); La. Rev. Stat. Ann. sec. 17: 286.3 (1981). For a complete discussion of these statutes and the litigation that they spawned, see Edward J. Larson, *Trial and Error: The American Controversy Over Creation and Evolution* (New York: Oxford University Press, 1989), 125–88.

29. "Remember Scopes Trial? ACLU Does," *Times-Picayune* (New Orleans), 22 July 1981, p. 1.

30. Aguillard v. Edwards, 765 F.2d 1251, 1253 and 1257 (5th Cir. 1985); Aguillard v. Edwards, 778 F.2d 225, 226 (5th Cir. 1985)(Gee, J., dissenting).

31. Edwards v. Aguillard, 482 U.S. 578, 590, 590 n. 10 (1986); ibid. at 603 (Powell, J., concurring); ibid. at 638 (Scalia, J., dissenting)(citation omitted).

32. Stephen L. Carter, *The Culture of Disbelief: How American Law and Politics Trivialize Religious Devotion* (New York: Basic Books, 1993), 169, 175–76, 178.

33. Foreword, in Henry M. Morris, ed., *Scientific Creationism*, gen. ed. (San Diego: Creation-Life, 1974), iii, v.

34. Bernard Ramm, *The Christian View of Science and Scripture* (Grand Rapids, Mich.: Eerdmans, 1954), 260. Regarding Graham's endorsement, see Ronald L. Numbers, *The Creationists: The Evolution of Scientific Creationism* (New York: Knopf, 1992), 185.

35. Tom Curley, "New Life in Evolution Debate," *USA Today*, 27 March 1996, p. 3A; Millicent Lawton, "Ala. Board Mulls Taking Stand on Evolution as Theory," *Education Week*, 8 November 1995, p. 13.

36. Peter Applebome, "70 Years After Scopes Trial, Creation Debate Lives," *New York Times*, 10 March 1996, p. 1.

37. Paula Wade, "Attempt to Amend 'Monkey Bill' Revives Debate over Darwin, God and Teacher," *Commercial Appeal* (Memphis), 5 March 1996, p. A1.

38. Andy Sher and Alison LaPolt, "Senators Slap Hold on 'Son of Scopes' Bill; Sponsor Vows Return," *Nashville Banner*, 5 March 1996, p. B4; "Echoes of Scopes Trial," *Nashville Banner*, 4 March 1996, p. A10; "Evolution Bill Makes It Through House Panel," *Jackson Sun* (Tenn.), 28 February 1996, p. 10A; Andy Sher, "Evolution-Bill Opponents Toss in Monkey Wrenches," *Nashville Banner*, 4 March 1996, p. B2; Vicki Brown, "Evolution Bill Killed in Senate," *Cookeville Herald-Citizen*, 29 March 1996, p. 2.

39. Dan George, "60 Years After Scopes, Town Is Much the Same," *Indianapolis Star*, 21 June 1985, p. 19A; Jane DeBose, "New Battle Over Evolution," *Atlanta Constitution*, 6 March 1996, p. B5; Ann LoLordo, "Tennessee Legislature Might Try Scopes Again," *Baltimore Sun*, 7 March 1996, p. 1A.

40. "Inherit the Wind: The State Sequel," *Nashville Tennessean*, 3 March 1996, p. 6A.

41. Lewis Funke, "'Inherit the Wind' Is Play upon History," *New York Times*, 22 April 1955, p. 20; Vincent Canby, "Of Monkeys, Reason and the Creation," *New York Times*, 5 April 1996, p. C1; "Mixed Bag," *The New Yorker*, 30 April 1955, p. 67; "The Theater," *The New Yorker*, 22 April 1996, p. 12; Roger Rosenblatt, "Inherit the Wind," *News Hour with Jim Lehrer*, 13 May 1996.

42. Richard Hofstadter, *Anti-Intellectualism in American Life* (New York: Knopf, 1963), 129.

43. Tony Randall, personal communication, April 1996.

44. Canby, "Of Monkeys, Reason and Creation," C1.

INDEX